GEOGRAPHY
GCSE AQA

GEOGRAPHY GCSE AQA

Human and Physical Geography

Elizabeth Paice

Copyright

Dedication

For my family

Table of Contents

Introduction

The study of geography is very much about what has happened, what is happening and what might happen in the future. As Winston Churchill once said: "The farther back you can look, the farther forward you are likely to see." Geographers look around them where ever they are – whether it is in the town or country – and they make observations and measurements. It is by using this data that geographers can make judgements, draw conclusions and offer evaluations.

The structure of the book follows the AQA GCSE GEOGRAPHY (8035) Specification for teaching from September 2016 onwards and for exams in 2018 onwards. There is comprehensive coverage of all the topics specified.

This textbook has been written using information from textbooks, examination mark schemes and the internet. Where ever possible the facts and figures come from UK government websites, UK based organisations such as the BBC, and reputable international organisations including the United Nations and the World Bank. All measurements are approximate.

As the world is a dynamic place, no doubt some of the information in this textbook will date very quickly. If there are any errors, please do notify me of them so that I can make corrections for future editions.

Elizabeth Paice
London, September 2023

1. Living with the physical environment

The Earth is a dynamic planet, which is constantly changing, as complex processes and systems shape the physical environment. The world's many distinct landforms and landscapes such as plains, mountains, valleys and deserts, have formed as a result of these changes and connections. Geomorphology is the study of the physical features of the Earth and its crust, and their relationship with the rocks which lie beneath. An understanding and knowledge of geomorphology helps people live with their physical environment.

Within the Earth's crust, the movement of the plates and other tectonic processes leads to the formation of volcanoes and earthquakes. On the Earth's surface, coastal and fluvial processes such as weathering, erosion, deposition and transportation alter the shape of rivers, coasts and glaciated highlands.

In the planet's many different ecosystems, living organisms interact with each other and the physical environment in order to create and sustain life. The Earth has a rich biodiversity – an enormous variety of life. Biodiversity includes all the wildlife, plants, fungi and bacteria which depend on biological processes such as photosynthesis, and systems like the nutrient cycle.

In the atmosphere, the global circulation of the air helps to determine the world's patterns of weather and climate. Meteorological processes including evaporation and condensation also influence the condition of the atmosphere. As a result of the different climatic zones, unique environments such as tropical rainforests, hot deserts and cold environments form in specific places.

Natural hazards are extreme events which are caused by the forces of nature. These dangers pose major risks to people and property because they can bring death, damage and destruction. Tectonic hazards include earthquakes, volcanic eruptions and tsunamis. Weather hazards encompass tropical storms, floods, droughts and heatwaves.

Humans interact with both the physical environment and the natural world. Human activities such as farming, building cities, burning fossil fuels and damming rivers are changing the world about us in many ways at local, national and international scales.

Climate change is the large-scale, long-term shift in the average temperatures and weather patterns of the Earth. As a result of human activities, the amount of greenhouse gases in the atmosphere is increasing. Greenhouse gases, which include carbon dioxide, methane and nitrous oxide, contribute to the greenhouse effect. The greenhouse effect is like a blanket around the Earth which prevents the Sun's heat from escaping back to space. This results in global warming which is a rise in the Earth's surface temperature. As a result, more extreme weather events are occurring.

The UK has a range of diverse landscapes which include uplands, lowlands and river systems. There are coastal, fluvial and glacial landscapes with their associated landforms such as bays and headlands, valleys and flood plains, corries and ribbon lakes.

Geography GCSE AQA

An understanding of the direct and indirect effects of human interactions on the physical environment and the living world, can help governments choose the right management strategies to ensure that life on Earth is sustainable for future generations.

2. The challenge of natural hazards

A definition of a natural hazard is that it is an extreme event caused by the forces of nature, which pose a major threat to human life. The likelihood that a powerful earthquake or a destructive hurricane may occur is a risk that many people take when deciding where to live or work. Natural hazards not only can kill humans and wildlife, but they also can disrupt everyday life and destroy infrastructure and ecosystems. Types of natural hazard include tectonic and weather events.

Tectonic events such as earthquakes and volcanoes are natural hazards. These dangerous occurrences are caused by the movement of the tectonic plates. Earthquakes, which are sudden or violent movements within the Earth's crust followed by a series of shocks, are one type of tectonic hazard. Volcanic eruptions occur when volcanoes, which are openings in the Earth's crust, erupt spewing out lava, ash and gases.

Extreme weather events are also natural hazards. Examples of extreme weather include severe snow blizzards, heat waves, floods, droughts and tropical storms. These hazards are significantly different from the average or usual weather because they are especially severe and maybe unseasonal. The length of an extreme weather event can vary. Sometimes the event only lasts a few hours but at other times they may persist for many days or weeks.

In the tropics, extreme weather events such as tropical storms, which are also called hurricanes, cyclones or typhoons, can be particularly dangerous. These areas of low pressure, which are accompanied by heavy rainfall and powerful winds spiralling around the calm central point – the eye of the storm – can cause loss of life and damage to property.

The probability of a natural hazard occurring is not the same everywhere – some parts of the world are more at risk of experiencing a major natural hazard than others. For example, Japan and the other countries located along the Ring of Fire at the edge of the Pacific Plate, have a higher chance of experiencing frequent large earthquakes than other places. Tectonic activity is rare in countries such as the UK, which are located far from a plate boundary.

When a natural hazard occurs, the immediate responses are the reactions of people during and straight after the disaster. The long-term responses are the actions that people take in the weeks, months and years following the event.

In order to forecast when a natural hazard may occur, scientists monitor and record any physical changes to the ground or the atmosphere. For example, they may measure the earthquake tremors around a volcano or use satellite photographs to track the path of a tropical storm.

Steps can be taken to protect people and property before a hazard strikes to reduce its impact. These actions can include educating people or improving building design. Communities also prepare action plans which enable them to respond to, and recover from, natural disasters.

These plans can include preparations to evacuate the area, to broadcast information, or to sound warning sirens.

Meteorologists and geologists collect data in order to be able to predict and attempt to forecast when and where a natural hazard will strike. Volcanic eruptions and tropical storms are easier to predict than an earthquake, but even so, forecasting when and where a natural hazard will occur is not an exact science.

The primary effects of a natural hazard are the initial impacts of the event on people and property. For example, during and after an earthquake or a tropical storm, buildings may collapse or be blown away.

The secondary effects are the incidents that occur sometimes much later as an indirect result of a natural event. For instance, earthquakes can rupture gas mains causing fires to break out. Tropical storms can wash pollutants such as raw sewage into rivers and reservoirs, thus reducing supplies of clean water and causing outbreaks of disease.

The economic impacts of a natural hazard are the effects on the wealth of an area or community. The social impacts are the effects on the lives of the people living in the affected area. The effects of the natural hazard on the landscape and ecology are the environmental impacts. Management strategies are the techniques which can be used to control, respond to, or deal with an event.

There are environmental, economic, social and technological factors which affect hazard risk. Physical environmental factors include the size, intensity, duration and magnitude of a natural event. A large volcanic eruption, which continues for a long time, will cause more damage and destruction than a small short eruption. Equally, light rain showers rarely cause any damage, but days of heavy rainfall can result in widespread flooding.

Some natural hazards occur in specific areas. For example, tropical storms form over the Pacific, Atlantic and Indian Oceans and are clustered in tropical areas such as the Philippines in south-east Asia. Flooding however is a natural hazard that can happen anywhere. Although low lying countries with large floodplains, such as Bangladesh, are more likely to experience large-scale flooding than countries with lower rainfall and higher relief,

The risk of a major natural hazard occurring can vary even within an area. For example, a dormant volcano which shows few signs of activity poses a low risk. In contrast, an active volcano which is having frequent small eruptions and is growing in size, is more dangerous.

Economic factors include the wealth of a country. A higher-income country (HIC) can afford to invest in higher quality infrastructure and buildings, which are better able to withstand the forces of nature. In addition, HICs have hazard warning systems and emergency services, which prevent the loss of life. In contrast, a low-income country may be less well prepared for a natural disaster, with poorly constructed buildings, and smaller search and rescue teams resulting in a higher number of deaths.

Social factors include the level of development and the number of people living in an area. If an earthquake hits a densely populated place like a city, the death rate is likely to be higher than if a quake occurs in a sparsely populated desert or rural area. If a country has good emergency services and many doctors and hospitals, the impact of natural hazards can be lower, because the response to the disaster can be better, resulting in fewer deaths and injuries.

Technological factors include having the science and technology to be able to predict more accurately when a natural disaster is about to occur. Vulcanologists, seismologists and meteorologists use historic records, and computers and scientific equipment to collect, record and analyse data. This information can help them to forecast where a major natural hazard is likely to happen, although not necessarily when.

3. Tectonic hazards

Earthquakes and volcanic eruptions are the result of physical processes. Plate tectonics theory explains why earthquakes and volcanoes occur and how new mountains are formed. The theory first emerged in the early twentieth century when Alfred Lothar Wegener, a German scientist, proposed the idea of continental drift. Wegener suggested that there was once a supercontinent, that had broken up into smaller parts which had moved apart. In an atlas you will see that South America and Africa look as if they were once joined together. Later, scientists studying the world's oceans, found more evidence to support this idea. Today the theory that the Earth's surface is divided into a series of tectonic plates along the edges of which land is being created and destroyed is generally accepted. A plate has now been identified as being a rigid segment of the Earth's crust which can float across the heavier, semi molten rock below.

The Earth is composed of different layers, like an onion. At the centre of the Earth is a solid metal inner core and around it a molten metal outer core. The next layer is called the mantle. The mantle is very thick and is composed of rocks some of which are slow moving and semi-molten. Above the mantle and nearest to the Earth's surface is the crust which is also made of rock.

The world's plates are found in the top layer of the Earth and are composed of the crust and part of the mantle. The plates move very slowly – only a few centimetres every year – either towards, away from or alongside each other. This movement is powered by the convection currents in the mantle. These convection currents move in big ovals and are driven by the hot magma rising and the cold magma sinking. There are seven major plates and six smaller ones. The movement of these plates causes the formation of earthquakes and volcanoes.

Under the sea is a layer of oceanic crust which is 10 kms deep and is made of dense igneous rocks. An oceanic plate is thinner and heavier than a continental plate. Below the land is a thick layer of continental crust. This layer is seventy kilometres thick and is made up of sedimentary, metamorphic and igneous rocks. A continental plate is less dense, lighter and thicker than an oceanic plate. A plate margin is the boundary between two tectonic plates.

The global distribution of earthquakes and volcanic activity is uneven. Most earthquakes and volcanoes occur in belts, which are long narrow zones along plate margins. The largest earthquake belt runs along the edge of the Pacific Plate, which lies under the Pacific Ocean and is the world's largest plate. Over 80% of large earthquakes occur in this cluster, which is known as the Ring of Fire. Many places are located in the Ring of Fire including the Kamchatka Peninsula, Alaska, Japan, the Andes, New Zealand and Mexico.

Other earthquake belts are found along the Mid-Atlantic Ridge in the middle of the Atlantic Ocean; in North America at the conservative plate margin of the North American Plate and the Pacific Plate; and along the destructive plate margin in South America, where the Nazca Plate meets the South American Plate.

Earthquakes sometimes occur far from a plate boundary as the result of human activity or old rocks moving into a new position. Although far from the Mid-Atlantic Ridge, which is the nearest plate margin, the UK has about 250 earthquakes a year, although most of these are very small.

Volcanoes also occur in linear clusters along the plate margins. For example, the Ring of Fire is the most volcanically active zone in the world. Mount Fuji in Japan, Krakatoa in Java, and Mauna Loa in Hawaii are just some of the dangerous volcanoes located in the Ring of Fire.

There is a clear relationship between earthquakes and volcanic eruptions and plate margins. In other words, most earthquakes and all volcanic eruptions occur along the plate margins. At the plate boundaries the tectonic plates move towards, away or past each other. This movement causes pressure and tension to build up and cracks and fault lines in the Earth's crust to develop. Earthquakes occur when this pressure and tension is suddenly released. Volcanoes occur in places where hot magma rises up from the mantle below into the gaps left when the tectonic plates move apart. Volcanic activity also happens where the plates are converging and crust is being subducted.

At the plate margins, there are physical processes powered by the convection currents. These processes destroy old crust, cause earthquakes and volcanic activity, and create new crust and landforms. There are three types of plate margin – constructive, destructive and conservative. The processes taking place are different at each one.

At a constructive plate margin, the tectonic plates are diverging moving apart leaving a gap through which rising magma adds new material to the plates. The Mid-Atlantic Ridge is an example of a constructive plate margin. It runs down the middle of the Atlantic Ocean and is the meeting point of the Eurasian and African plates with the North and South American Plates.

Hot magma from the mantle rises up the gap left by the plates moving apart. If the magma carries on rising through the crust it can reach the Earth's surface and run over the land in a gentle eruption. Hot runny magma can flow a long way and can form a shield volcano with gently sloping sides, for example Skjaldbreiður in Iceland on the Mid-Atlantic Ridge.

When magma rises to the crust's surface under the sea, the pressure and tension causes cracks to develop in the ocean floor. The hot magma rises up through these fault lines and creates mid-ocean ridges which become higher and higher. As the magma cools, it becomes denser and slides down the side of the ridge. This causes the other plates to move away from each other in a movement called ridge push.

Earthquakes also occur along the fault lines found at constructive plate margins. They are usually of low magnitude, although some can be large and linked to volcanic activity. Most happen at shallow depths below the surface where the plates are moving apart.

At a destructive plate margin, the plates are converging or moving together and oceanic plate is subducted. It can be associated with violent earthquakes and explosive volcanoes. If a continental plate moves towards an oceanic one, the oceanic plate is forced down into the mantle where it is destroyed in the deep subduction zone. Violent volcanic eruptions sometimes occur when this happens. If the magma is thick and sticky, it forms steep sided composite volcanoes. An example of a destructive plate margin can be found in the Andes along the west coast of South America where the Nazca and South American plates are converging. San Pedro is a composite volcano in northern Chile and one of the tallest active volcanoes in the world.

Oceanic trenches are also formed in subduction zones. When a heavy oceanic plate slides down into the mantle, it moves under a lighter continental plate, in a process called slab pull. This causes the sea floor to bend into an oceanic trench, which is a steep 'V' shaped depression.

If two continental plates collide, both plates are pushed upwards, forming new fold mountains such as the Himalayas.

At a conservative plate margin, the tectonic plates are sliding sideways past each other. As they move their jagged edges get jammed together, causing a build-up of pressure and tension along the various fault lines. If one plate is moving faster than the other, this can cause the pressure and tension to build up even further. Sometimes this pressure is released as an earthquake. At a conservative plate margin crust isn't created or destroyed but the shape of the landscape can be changed. These plate margins are also called passive or transform boundaries. The San Andreas fault line in California is a conservative plate margin.

The effects of, and responses to, a tectonic hazard vary between areas of contrasting levels of wealth. Broadly speaking, poorer countries experience worse effects and poorer responses than richer countries.

By studying different examples of volcanic eruptions and earthquakes, we can understand more about the primary and secondary effects of tectonic hazards and the immediate and long-term responses to them.

After being dormant for centuries, Mount Vesuvius near Naples in southern Italy, erupted in AD 79 taking the residents of Pompeii and Herculaneum by surprise. The force of the eruption led to a huge cloud of volcanic ash, gas and pumice stones the size of cannonballs, being blown ten miles into the sky. The eruption continued for hours, showering Pompeii with red hot ash and cinders. A pyroclastic flow of ash and pumice and a cloud of deadly gas covered the immediate area.

Herculaneum, a city of five thousand people on the coast, initially escaped devastation but then a dark horrible cloud of very hot ash and gas surged down the west slopes of Vesuvius killing every living thing in its path. A river of volcanic mud and rock which followed covered the city in a thick layer of mud and ash.

The primary effects of the eruption, which lasted two days, were to bury both cities under thick layers of volcanic material. Many people and animals lost their lives and there was widescale destruction of property.

The secondary effects were that both Pompeii and Herculaneum were destroyed and then abandoned as over five thousand people had been killed and many more injured and left homeless. The nearby coastline was altered and the surrounding countryside covered in thick layers of ash. There were food shortages as crops had been destroyed and farming took decades to recover.

The immediate response of most people to the eruption of Vesuvius was to flee for their lives although some found temporary refuge in cellars and under stone tables before being overcome. In the region a search and rescue operation swung into action. Boats set sail from Naples to save people and take them to safety.

Long-term responses occur over weeks, months or years. When memories of the eruption had faded and the ash had weathered into a fertile soil people returned to their farms and villages. Pompeii and Herculaneum lay buried with the dead entombed until excavations began in the last century. Today Mount Vesuvius and Pompeii are popular tourist attractions bringing in much needed income to the Naples area. The fertile farmland around Naples produces olives, vines, nuts and fruit such as oranges and lemons, which grow on the volcanic soils surrounding Mount Vesuvius.

Mount St. Helens is a composite volcano 2,549 metres high, located in the Cascade Range in western North America. In 1980 an eruption of the volcano began with a series of earth tremors along the north side which triggered a small volcanic eruption. Steam and ash were emitted through the volcano's crater and vents.

Small eruptions continued for weeks during which the mountain began to change shape as a large bulge made up of magma moved up inside the northern side of the mountain. Then a larger earthquake, of 5.0 magnitude on the Richter scale, triggered a bigger eruption. The north side of the volcano collapsed sending an avalanche of rock and ice down the mountain causing Spirit Lake to flood. Thirteen miles of the Toutle valley were buried in deep layers of rocks and mud.

A lateral blast of steam and hot gas sped along the ground destroying everything in its path including ten million trees. Another eruption sent a mushroom shaped cloud of gas and hot ash twelve miles into the sky.

The main primary effects of the eruption were on the physical landscape which had been completely changed. Nearly 60 metres of the top of Mount St. Helens had been blasted away leaving a horseshoe-shaped crater. The pyroclastic flows had flattened forests, blocked roads, damaged bridges and killed people and animals. The mudflows and floods devastated the countryside. As Mount St. Helens was located in a sparsely populated national park, casualties were relatively low. Although the evacuation of the area had started straight away, some refused to leave resulting in 57 deaths.

The secondary effects were that normal life was disrupted for months. For two weeks ash rained down on the surrounding countryside and settlements before drifting around the globe. The clean-up operation and repairs cost millions. It took years for nature to recover fully.

The immediate response to the eruption was to search for survivors. Helicopters flew over the affected areas looking for signs of life. People stranded in their homes or cars were found, taken to safety and given first aid. Roads blocked by ash and rocks were cleared. Roads, bridges and buildings were rebuilt. Anyone made homeless was given temporary shelter until their homes were safe to return to. Face masks were distributed to people until the ash had blown away.

The long-term responses to the eruption of Mount St. Helens, included clearing the burnt vegetation and planting young trees. The ash from the eruption had enriched the soil enabling the new forests to grow quickly. When nature and wildlife had recovered the tourists returned, boosting the local economy. Scientists have continued to monitor the volcano with the latest equipment.

In January 2010, Haiti, a lower-income country in the Caribbean, was hit by a 7.0 magnitude earthquake. The epicentre was 16 miles away from the capital Port-au-Prince, which sits on an active fault line. In the following two weeks there were sixty aftershocks.

The primary effects of the earthquake were catastrophic, leaving Port-au-Prince in a state of devastation. Thousands of people were killed and injured by falling masonry, collapsing walls and flying glass. Others fled for their lives. Not only were many homes destroyed but also vital infrastructure such as government offices, hospitals and the airport. Many roads were blocked and the port was damaged. In addition, the telecommunications system and the electricity supply were cut.

The secondary effects of the 2010 earthquake in Haiti are still evident today. The economy took a long time to recover as businesses and trade were disrupted due to the damage to factories and shops, and the shortages of workers and power. The standards of living and the quality of life of many people recovered slowly as homes and hospitals had to be rebuilt, supplies of clean water restored, and public services resumed. Homelessness, outbreaks of disease – such as cholera – and civil disorder are problems that have not been fully solved.

The immediate responses to the earthquake were slow and disorganised as Haiti was ill-prepared for an earthquake. Many firefighters, police officers and civil servants had been killed in the quake. The damaged infrastructure made it difficult to reach the people and places needing help. Unable to cope on their own, the Haitians requested international aid. The Dominican Republic, Haiti's neighbour, immediately accepted refugees and sent rescuers. Many other countries, including Britain, sent search and rescue teams equipped with sniffer dogs and heavy lifting equipment. Survivors needing medical treatment were airlifted on to American hospital ships. The dead were buried, sometimes in mass graves. The homeless were housed in temporary camps. Aid supplies were distributed.

The long-term response to the earthquake included rebuilding the country's infrastructure as well as the capital city Port-au-Prince, and a new port. This was financed by aid money and loans from the World Bank. Haiti is now better prepared for another earthquake as many homes and buildings have been earthquake proofed.

One of the countries most prepared for an earthquake is Japan, a high-income country in East Asia. On 11 March 2011, a 9.0 magnitude earthquake shook the north-east of Japan for 6 minutes. The epicentre of the earthquake was located in the Pacific Ocean, 72 kms east of Tohoku, at a depth of 24 km. It occurred in a subduction zone where the Pacific plate dives under the Eurasian plate. The shock waves were felt across northern Japan and in Tokyo. Hundreds of aftershocks occurred in the following weeks.

The primary effect of the earthquake was the formation of a series of tsunami waves caused by the sudden movement of the plate margin displacing the sea above. When the 40 metres high tsunami waves hit the Japanese coast, they breached high sea walls and flowed 10 km inland as far as Sendai, flooding a widespread area. The tsunami also travelled across the Pacific to Alaska, Hawaii and Chile. The power of the waves was so great that everything in their path was smashed to pieces. Many people were drowned immediately. It is estimated that 20,000 people were killed. Many are still missing. In spite of Japan's strict building code, more than half a million homes and buildings were damaged or destroyed. The infrastructure of north-east Japan was crippled.

The secondary effects of the earthquake and tsunami were that everyday life was severely disrupted for weeks due to the scale and strength of the waves. There were temporary food shortages and power cuts. Many people were traumatised by the disaster. However, the biggest crisis was the effect of the tsunami on the Fukushima Daiichi Nuclear Power Plant. The flooding triggered a nuclear meltdown and the release of radioactive materials into the environment.

The immediate response to the earthquake and tsunami was the triggering of Japan's early warning system which sent text messages to mobile phones. Trains and factories came to a halt as people sought shelter. In Tokyo the residents only had a minute to take refuge before the shock waves hit the city. The emergency services then swung into action. Survivors were rescued, taken to safety or given appropriate medical help. Whole communities were moved to inland higher ground in case there was another tsunami. Clean water, food and shelter were provided to those made homeless and separated from their families. The dead were found amongst the debris and buried.

The long-term response to the tsunami in Japan is still taking place. The affected areas have been rebuilt at a cost of $200 billion making it the costliest natural disaster in world history. Buildings and infrastructure have been redesigned and rebuilt using different materials to make them more earthquake proof. The Fukushima Nuclear Power Plant has been closed and the surrounding land decontaminated. However, the recovery operation created jobs for builders, electricians and carpenters. Japan has been able to recover far more quickly than a poorer country.

Why do people live in places at risk from a tectonic hazard such as a volcanic eruption, earthquake or tsunami? There are many physical, economic, social and technological reasons.

The physical environment around volcanoes can be favourable places to live. When a volcano erupts, the lava and ash emitted covers the surrounding land and over time weathers into a fertile soil. If the lava flows reach the coast, they cool and harden into new land which can be used to build new towns and ports. This is happening on Hawaii.

In spite of these advantages, living near to a volcano can be dangerous. If an eruption occurs there is a risk to life and livelihoods, and the destruction of both the built and the natural environment.

An advantage to living along a fault line is the presence of fresh water. Desert Hot Springs in California is built on top of the San Andreas fault line. Its water supply comes from underground enabling a city to be built in the desert. Of course, the disadvantage of living near a fault line is the threat of an earthquake which could result in death and destruction.

There are good economic reasons to live near a volcano. Farming is a major activity as the volcanic soils are full of natural fertilizers such as magnesium, iron, phosphorus and potassium, and are porous, which means they retain moisture for longer. Crop yields are higher in these rich soils which makes the land good for farming. Several types of ash are deposited during an eruption, and these weather into different types of soil enabling a wider variety of crops to be grown. When Mount Nyiragongo in eastern Congo erupted in 2002, the countryside was flattened by flowing lava. However, afterwards there were better harvests of bananas and cabbages which led to higher incomes for the farmers.

Tourism is also another economic activity found in volcanic areas. Tourists visit the volcanoes, hot springs and geysers. The local economy benefits from this economic activity. There are more jobs and opportunities as well as increased tax revenues. The tourist industry in Iceland, which has a lot of tectonic activity, accounts for 38% of its exports and employs 15% of its workforce. One of its top tourist attractions is the Blue Lagoon, a geothermal spa which has 1.5 million visitors per year generating a lot of income.

Volcanic areas are rich in resources. During periods of tectonic activity, rocks are changed into igneous rocks such as basalt which are used in construction and road building. During eruptions, gems, minerals and metals such as gold, silver, diamonds, copper and zinc can be formed by the heat and pressure.

In volcanic areas, hot water comes to the surface in the form of hot springs and geysers. This constant supply of very hot water is used to heat homes, greenhouses and swimming pools in many places such as Iceland.

Geothermal energy can also be produced from the boiling water heated by the magma and trapped underground beneath a dense layer of impermeable rock. A deep hole is drilled down to collect the steam which rises to the surface and expands, either driving turbines or being directed through heat exchangers. Geothermal power stations in Iceland and New Zealand produce electricity from this renewable source.

However, the disadvantage of living in a tectonic area is that economic activity can be slowed or halted by an eruption or earthquake. The damage caused to homes, businesses and communication systems and other vital infrastructure can cause a loss of income and result in high rebuilding costs.

There are also many social reasons why people live in tectonic areas. Attractive places to live such as along the coast are often on top of a plate boundary. For example, in California, many people live in San Francisco and Los Angeles because of the jobs and opportunities on offer. However, these cities lie on the San Andreas fault line and there is the risk of an earthquake. People choose to stay because they are settled and they cannot afford to leave their jobs and communities to make their lives in a safer place.

Most of the people who live in tectonic areas believe that the chances of an eruption or an earthquake are slim. They rely on the early warning systems giving them time to find refuge. They hope that the emergency services will rescue them and that the government will help them rebuild afterwards. A few people are simply not aware that a risk of an eruption or an earthquake exists.

There are also technological reasons why people stay in a tectonic area. Engineering can make people feel safe. Buildings can be constructed to be earthquake proof. Local and national governments can be well prepared for a possible earthquake or volcanic eruption with evacuation plans, early warning systems and advice. These preparations can reduce the risk but disaster can overwhelm even the best prepared places.

How do countries such as Japan, Iceland and New Zealand prepare for volcanic eruptions and earthquakes? They use monitoring, prediction, protection and planning.

First of all, scientists monitor volcanoes and fault lines carefully. They set up stations over a wide area to observe and measure any changes on the Earth's surface which could indicate that magma is rising in a volcano or pressure and tension increasing along a fault line. They use remote sensing, laser beams and tiltmeters to detect any changes to ground temperature and the shape of the land. Seismographs record any earth tremors and other instruments detect changes to the number of different gases in the air and hot springs. Other warning signs that a tectonic event is imminent include melting snow, swellings on the sides of volcanoes and cracks in the ground.

All the data is collected, recorded and analysed to see if any patterns are emerging. However, it is easier to predict where a volcanic eruption or earthquake will happen, than to predict when it will occur. Even with monitoring data, observations of changes to animal behaviour, and historical records of previous eruptions or earthquakes, people can still be taken by surprise.

If scientists think a volcano is about to erupt, they can warn people so that they have enough time to leave their homes and go to a place of safety. This happened in Iceland when scientists correctly forecast the eruption of Eyjafjallajökull in 2010. However, if there are too many false alarms, people may ignore the warnings.

Engineers can also try to protect people from the worst effects of an eruption or earthquake. On the ground, barriers can be made to divert lava flows away from homes – this has happened on the slopes of Mount Etna in Italy where earth embankments have been erected. In Japan, tsunami walls along the coast have been built to protect homes and important buildings like nuclear power stations.

Another way of protecting people is to ban the building of homes in areas most at risk. Hazard maps showing the places most likely to be affected by a volcanic eruption or earthquake can be drawn up and used to deny or grant planning permission. Only new buildings which can withstand being shaken during an earthquake would be permitted.

To earthquake proof buildings, rubber shock absorbers can be installed in the foundations to absorb the tremors, and rolling weights put on roofs to steady the building during a quake. To prevent fires and accidents from falling glass, automatic switches to turn off the gas and electricity, and to close window shutters can be fitted. Buildings can be prevented from collapsing by reinforcing walls with steel and concrete and wrapping pillars in steel frames to allow them to sway safely during an earthquake.

Governments can plan what to do in the event of a volcanic eruption or an earthquake. An early warning system can be installed that sounds a siren or sends text messages to people's phones. Emergency messages can be broadcast on television and the radio. Information leaflets can be sent to homes and organisations. Families, schools and businesses can hold regular evacuation drills so people know what to do and where to go if disaster strikes.

The emergency services can also practice their responses in the event of a tectonic event. People can be evacuated to nearby empty spaces such as parks or to community centres. Here people can be kept safe and given aid such as temporary shelter and medical attention. Stockpiles of tents, blankets, water, food and medicines can be kept in case of such an emergency. These can be moved to safer areas if need be.

After a volcanic eruption or an earthquake, a well-organised search and rescue operation needs to be mounted to find trapped people and to bury the dead. There should be a co-ordinated clear up operation to reopen roads, repair bridges, reconnect gas and electricity and move the debris. A re-building and re-planting programme should follow. A well-planned response to the disaster will enable life in the immediate area to get back to normal as quickly as possible.

4. Weather hazards

Global atmospheric circulation, the worldwide system of winds, helps to determine the world's pattern of weather and climate. The general atmospheric circulation model shows how the globe's pressure belts and surface winds transport the equatorial heat from the lower latitudes to the cooler parts of the world at the poles.

In each hemisphere air circulates through the entire depth of the troposphere, which extends up to 15 km. The Sun's heat powers the circulation. The shorter the distance that the Sun's rays travel to reach the Earth's surface, the hotter they remain. The hottest place on Earth is along the Equator where the Sun is directly overhead all the time.

The Earth has belts of high and low air pressure which encircle the globe. High pressure belts occur where the air is sinking, and low-pressure belts where the air is rising. The movement of the air within the atmosphere dictates the weather, which can become extreme at times.

At the Equator there is an equatorial low where the hot air is rising. This allows the winds to blow in. The climate here is very hot and wet all year round. There are no seasons and there is a small annual temperature range. Every day the morning sun heats the air which rises into the atmosphere where it cools, condenses and forms clouds leading to heavy rain showers and thunderstorms in the afternoon. In these humid conditions, the equatorial rainforests grow quickly. Countries which have equatorial climates include the Democratic Republic of the Congo, and Indonesia.

The equatorial air is blown towards latitudes 30°N and 30°S, where it sinks and forms belts of high pressure called sub-tropical highs. In these places, daytime temperatures are very high but during the night it is very cold. The high pressure prevents rain bearing winds from blowing in, so the climate is very dry. Most of the world's hot deserts including the Sahara Desert are located in these sub-tropical high-pressure belts.

At mid-latitudes 60°N and 60°S there are low pressure belts called the sub-polar lows. Here the warm air from the sub-tropics meets cold air from the poles. As the warmer air is less dense than the cold air, it rises creating low pressure. The weather here is unstable with rain and wind. The UK is located at 55°N which explains why the weather in the UK is often unsettled and changeable.

At the North Pole 90°N and the South Pole 90°S there is very cold sinking air. This creates belts of high pressure called the polar highs. A polar climate has cool summers and very cold winters and is found in Antarctica and Greenland.

The Earth has a clear pattern of surface winds which form part of the global atmospheric circulation. Winds are large scale movements of air caused by differences in air pressure. Winds blow from areas of high pressure to areas of low. This is because sinking air causes high pressure which leads to the winds moving away and diverging. In areas of low pressure, the winds meet or converge. Surface winds curve as they move because they are distorted by the Earth's rotation. They transfer heat and moisture from one place to another.

In the Northern Hemisphere the north-east trade winds blow south from the sub-tropical high-pressure belt at 30°N towards the equatorial low-pressure belt at the Equator. The warm moist westerlies blow north from the sub-tropical high-pressure belt at 30°N towards the sub-polar low-pressure area at 60°N. There they meet the very cold polar easterlies, which have blown south from the polar high-pressure belt.

Pressure belts and surface winds are affected by seasonal changes. In the summer when the Northern Hemisphere is tilted towards the Sun, the belts and the winds move north. In the winter they shift to the south. Both the type of air pressure and wind direction affects the weather. The south-westerly prevailing winds in the UK bring warm moist air and wet windy weather. Whereas the trade winds drive the formation of tropical storms.

Tropical storms are natural hazards that cause death and destruction. Known as hurricanes in the Caribbean and the USA, they are called cyclones in Asia and Australia, and typhoons in Japan and the Philippines. The tropical storm season lasts from June to October in Central America and the Caribbean; from December to March in the south-west Pacific, Madagascar and Australia; and from May to December in East Asia.

The global distribution of tropical storms (hurricanes, cyclones and typhoons) is uneven. They occur in clusters in the tropics between 5°N and 30°N, and 5°S and 30°S. They form over the Pacific, Atlantic and Indian Oceans and are most frequent in East Asia. They never form at the Equator or between Africa and South America.

The relationship between tropical storms and the general atmospheric circulation is a close one. Tropical storms only form into intense low pressure weather systems where the north-east and south-east trade winds are blowing. As these winds blow across the ocean, they start to spin in a process (called the Coriolis effect) which is caused by the rotation of the Earth. During a tropical storm the winds blow faster and faster until the air pressure rises and the storm dies out.

The causes of tropical storms are not fully understood. However, they develop over the ocean where three physical conditions are met. Firstly, the temperature of the sea should be warm – at least 27°C. Secondly, there should be low-level wind shear which means the wind should be changing speed and blowing in all different directions. The third requirement is that this location should be far enough from the equator for the winds to be able to spin or twist. These circumstances usually occur at the end of the summer or in the early autumn.

The sequence of the formation and development of a tropical storm is well documented. A storm starts where a lot of seawater is evaporating. The hot air rises, cools, condenses and forms massive cumulonimbus clouds. An area of low pressure develops and a strong wind blows in. Thunderstorms break out as there is a lot of energy being generated. The tropical storm starts to travel across the ocean powered by the evaporating water. It grows higher, wider and stronger as it does so. When the storm reaches land, it loses its energy and the winds slow down. However, the rain storms may carry on for days before stopping. Sometimes, the storm moves across the land and over another warm sea and picks up strength again.

The structure and features of a tropical storm are well documented. A tropical storm is an enormous mass of spinning white cloud, which can be over 300 miles wide and 5 miles high. At the centre of the storm is a circular cloudless calm area called the eye of the storm, which can be up to 30 miles wide. In the eye the cooler air sinks down to the ocean to be warmed up again. Surrounding the eye is a tall circular area of dense cloud called the eye wall. The eye wall is the most dangerous part of the storm because currents of warm air are rising rapidly and fanning outwards at the top. Here the strongest fastest winds, the heaviest rain and the most thunder and lightning are found. Beyond the eye wall the clouds are thinner and smaller, but the winds are still gusty, the rain heavy, and there is the occasional thunderstorm and tornado.

Tropical storms, which blow anti-clockwise in the northern hemisphere and clockwise in the southern, usually last for over a week. Accompanying these storms are high seas, storm surges, gale force winds of over 74 miles per hour (mph), torrential rain, flooding and tornadoes.

Climate change might affect the distribution, frequency and intensity of tropical storms especially if the average global temperature increases by 2°C as predicted. However, the distribution of tropical storms is unlikely to change that much. It is expected that during long hot summers, larger areas of the tropical oceans will reach 27°C and stay at that temperature for longer. This will increase the risk of tropical storm formation. It is possible that the storms will increase in frequency and intensity in the future. Also, as warmer air can hold more moisture, it is likely that hurricanes will be accompanied by heavier rain, leading to more coastal areas being flooded and damaged by the storms.

Tropical storms have significant effects on people and the environment. The negative impacts are greatest on places with low-lying coastal plains and high population densities. Low-income countries without coastal defences can be particularly badly hit by the extreme weather. Severe storms with very high wind speeds, torrential rain and large storm surges can cause considerable damage even to richer regions. Tropical storms can also affect the natural environment as the heavy rain can lead to widespread flooding, mud flows, and landslides. However, these seasonal storms do have a positive impact – the large amounts of fresh water they bring replenish the reservoirs of arid coastal areas.

The primary effect of a tropical storm is the death and destruction wrought by the gale-force winds and the torrential rain. Everything in the path of the storm can be damaged or destroyed. This includes homes, buildings and power lines. Trees can be uprooted, roofs blown off, and fields flooded. People can be killed or injured by collapsing buildings, flying debris, or drowned in swollen rivers.

The secondary effects become apparent when the storm has passed. The damage to homes, offices and factories can cause homelessness, disruption to everyday life and economic activity. The loss of power and communication systems can mean that households have no electricity, mobile phone signals or the internet. Flooding in the countryside can cause food shortages. Fires can break out if gas pipes have been severed.

The immediate response to a tropical storm is an assessment of the damage followed by the declaration of a state of emergency if necessary. The search and rescue operation usually starts immediately with sniffer dogs, helicopters and off-road vehicles to locate the dead and injured. People are taken to safety and given medical help. The dead are buried. The homeless are given shelter, hot food, blankets and medicines. If the affected country cannot cope on its own, the international community including governments and charities send in tents, bottled water, food parcels, field hospitals and any other necessary supplies.

The long-term responses to tropical storms include a clean-up and rebuild operation. Roads, bridges and coastal defences are rebuilt, as are homes – often to a higher standard and in safer places. Farmers are given assistance to replant their crops and to buy new animals. Aid agencies sometimes step in to replace items such as lost tractors and fishing boats. However, if large areas have been ruined and left covered in thick layers of mud, normal life may not resume for a long time.

A named example of a tropical storm is Typhoon Haiyan which hit central Philippines in 2013. This tropical storm had winds of over 195 mph and was one of the most powerful storms ever recorded. The storm destroyed many of the country's low-lying coastal farming and fishing communities, and displaced thousands of people.

The primary effects were that homes and other buildings were smashed into pieces by the wind. Leaves were torn from trees, fishing boats shattered, electricity lines brought down, crops destroyed and whole areas along the coast flooded by the storm surge. Over 6,000 people died and 4 million were made homeless. Power was cut off; the airport was badly damaged and roads were blocked by trees. In Tacloban, one of the main cities, the buildings were reduced to rubble and 4,000 people were killed.

Afterwards the secondary effects were social, economic and environmental. The people not only lost members of their families and their homes but often their livelihoods as well. Their health and education suffered as there were outbreaks of disease and many schools were closed. The economy was set back years as travel was disrupted and shops destroyed. Production in factories slowed and the loss of income was huge. Money had to be found to help clear up and rebuild, which was difficult as many people lacked insurance. Farmers lost their income as there was damage to rice and seed stocks. The combination of these problems led to an increase in food prices and looting. The environment was affected as well with a five-metre storm surge along the coast, which flooded forests and other ecosystems and contaminated land with pollutants.

The international community responded by providing food, water and shelter. The US sent search and rescue teams and emergency aid. The UK provided shelter kits containing a tent, blankets, mosquito nets, water containers, purifying tablets, tools and a solar light. People were evacuated to temporary camps until their homes had been rebuilt.

The long-term responses to Typhoon Haiyan have included attempts to make the country more resilient. New homes and cyclone shelters have been built away from areas at risk from coastal flooding. The government was given money to rebuild roads and bridges. Medical supplies were donated and farmers helped to replant rice and coconut trees. People were paid to clean up the debris and to rebuild buildings. The cost of rebuilding was estimated around $5.8 billion. A year on, thousands of people were living in temporary shelters and four million people were still displaced. Ten years later, life had almost returned to normal.

Monitoring, prediction, protection and planning can reduce the effects of tropical storms. To monitor the development of a storm, meteorologists use observation and measurement. Satellite images from space and data from weather stations, planes and ships at sea are analysed. Scientists based in hurricanes centres such as the one in Miami use special computer software to calculate the size of the storm.

To predict the path and strength of the storm the meteorologists use historic records of past storms and their computer models. To protect lives and property from tropical storms, storm proof buildings can be constructed. A hurricane proof building is reinforced with concrete, so that it can withstand strong winds, and has large gutters and drains to channel the heavy rain quickly away. Homes can also be built on stilts so that they don't get flooded. Strong sea defences and high sea walls can line the coast.

Planning for a tropical storm includes preparing coastal communities. People can be educated about the dangers with information leaflets, tropical storm awareness weeks and emergency drills. If a storm is forecast, warnings can be given and people evacuated to safer places. In Bangladesh these simple measures have saved many lives.

Although the UK does not experience tropical storms, it does have other weather hazards. In the winter there are strong winds, tornadoes and thunderstorms. During very cold weather, there can be periods of heavy snow, blizzards and hailstorms. Sometimes there are periods of prolonged rain leading to flooding. At other times there are droughts and heat waves. In spite of having a temperate maritime climate, the UK does have the occasional extreme weather event. This is unexpected, unusual or unseasonal weather, severe enough to pose a threat to life and property.

Extreme weather events can have several impacts on human activity. In the winter, heavy snow and ice can block roads, cover runways and damage overhead cables causing travel delays and cancellations. Remote communities can be isolated. Power cuts can occur. If the situation is prolonged, there can be a rise in the death rate. Frost and heavy snow can affect farming and may lead to food shortages.

During a summer heatwave, temperatures in the UK can exceed 40°C, which can be dangerous for the young and the elderly. As the UK infrastructure has not been designed to withstand extreme heat, travel can be disrupted by melting road surfaces and buckling train tracks. If there is a long period of hot dry weather, wildfires can breakout. During the 2022 heatwave, a wildfire devastated Washington a suburb in south-east London. Droughts can cause water shortages and necessitate the need for hosepipe bans. However, heatwaves can boost the tourism industry.

An example of a recent extreme weather event in the UK is the 2014 flooding of the Somerset Levels in south-west England. An area of 65 km² was flooded with 100 million m³ of floodwater for three months.

The causes of the flooding were that there was heavy rainfall over a long period. Water levels rose rapidly in the River Parrett and the other two rivers, which drain the low-lying flat area. The rivers burst their banks and floodwater spread over the farmland. There were high tides and storm surges in the Bristol Channel which prevented the floodwaters quickly draining away.

The social impacts included the disruption of daily lives. Homes and roads were flooded and villages such as Moorland cut off. Many residents temporarily moved out, and those who stayed used boats instead of cars. There were food shortages and power cuts.

The economic impacts were estimated at £10 million. Farming was disrupted. Businesses were affected by the closure of shops and the transport difficulties.

The environmental impacts were mainly caused by the floodwaters which contained a lot of sediment. The thin layer of alluvium left when the water receded enriched the soil. Unfortunately, there were incidents of pollution where chemicals and sewage contaminated the soil.

Various management strategies were used to reduce the risks to life and property. Before the floods occurred, the weather forecasters warned people that a period of prolonged heavy rain was expected and that flooding was likely. In response, people protected their homes and businesses with sandbags, moved valuables upstairs and stockpiled food, water and fuel. Farmers moved livestock to higher land.

During the floods, the emergency services pumped water out of buildings and off the flooded fields. Vulnerable people were looked after by community volunteers and given hot food or food parcels.

After the floods receded, the Somerset Rivers Authority dredged the rivers so that they could hold more water in future. The flood defences were repaired and improved. More pumping stations were built and mobile pumps bought. The height of river embankments, walls and roads were raised. A new tidal barrier across the River Parrett will prevent tidal flooding and protect 11,300 homes and 1,500 businesses in the future.

There is some evidence that the weather is becoming more extreme in the UK. Rainfall has increased in the last few decades. For example, the decade 2011 to 2020 was 9% wetter than 1961 to 1970. Six of the ten wettest years across the UK have occurred since 1998. Scotland has experienced the greatest increase in rainfall, while most southern and eastern areas of England have seen the least change. There is a trend towards an increase in the frequency and intensity of rainfall across the UK. However, we should expect some years to be wetter than others as this variation is a natural occurrence.

Temperatures have also increased. All of the UK's ten warmest years on record have occurred since 2002. Heatwaves are becoming more frequent. However, climate models suggest that winters will not become more extreme until the 2040s and summers not until the 2080s. Winters are predicted to become warmer and wetter. Summers are forecast to be hotter and drier. Summer rainfall is likely to be less frequent but more intense, causing severe flooding when it occurs. However, the UK will remain a temperate maritime climate with variable weather for the foreseeable future.

5. Climate change

Climate change is the large-scale, long-term change in the earth's climate. In recent decades there is compelling evidence that average temperatures have increased and the Earth's weather patterns have shifted. The term climate change can also be used to describe the various effects of an increase in greenhouse gases in the atmosphere, including more extreme weather events, changes in rainfall patterns, increased ocean acidification and rises in sea level.

Global warming is the effect of the rise in the Earth's surface temperature as a result of greenhouse gas emissions, which include carbon dioxide, methane and nitrous oxide. The greenhouse gases form an atmospheric blanket that stops the Sun's heat escaping back to space.

The change in the world's climate is the result of natural and human factors. The natural factors include changes in the orbit of the Earth, volcanic activity, and solar output. The use of fossil fuels, agriculture and deforestation are some of the human factors. The effects of climate change range from the risk of death from higher temperatures, to an increase in the number of areas at risk from flood, drought or wildfire.

The climate is a complex and dynamic system. Meteorologists use measurement and observation to collect and record a range of data about the weather and climate. At weather stations around the world, on land and at sea, daily temperature and rainfall totals are measured and recorded. This data along with satellite images and photographs is analysed by the scientists at the UK's Met Office using the latest computer software. To determine exactly what changes to the climate are happening, a very large data set from all over the world, needs to be built up and maintained for many years.

Scientists also study the climate of the Earth in the past. Evidence of what the climate used to be like can be found in ice cores, which are taken from the Arctic and Antarctic. The amounts of carbon dioxide and methane in the atmosphere over a period of time can be determined by an examination of the ice. Pollen in the sediment dredged from the bottom of lakes is examined in order to determine the types of plants that grew in the past. Evidence can also be found in tree rings, rocks and fossils.

The first recorded humans appeared and started to develop during the Quaternary period, which began 2.5 million years ago and which includes the Pleistocene and Holocene epochs. Climate change has been occurring since the beginning of this geological period as the Earth has experienced a sequence of glacial periods, followed by warmer interglacial millennia. In the last million years there have been at least ten ice ages. When the last glacial period ended 12,000 years ago, the Earth entered a warmer interglacial period. During this time there have been fluctuations in annual temperatures with some decades colder than others. So, to some extent climate change is a natural phenomenon.

However, there is evidence that climate change is now occurring at a faster rate than in previous millennia. The most significant change has been in the increase in global temperatures. After the last Ice Age ended, the average global temperature remained at around 14°C. It is thought the start of the rise in temperatures coincided with the beginning of the Industrial Revolution in 1750, and accelerated after 1950. Average global temperatures have now risen to around 15°C. The recent decade had some of the warmest years on record.

The rise in global temperatures has affected the sea. The upper layers of the oceans, down to a depth of 700 metres, are becoming warmer. Sea levels are also rising. Since satellite measurements began in 1993, the oceans have risen by 3 mm a year. The extent of sea ice is falling, especially in the Arctic.

The possible causes of climate change include natural factors. Orbital changes – the changes in the pathway of the Earth around the Sun could be responsible for climate change. The Earth orbits the Sun every year and is responsible for the seasons. Eccentricity, which is the shape of Earth's orbit, helps to determine our long-term climate. Changes in the shape of the orbit coincide with the timings of past glacial and interglacial periods. There is evidence that when the orbit is circular, the climate is warmer because the Earth receives more sunlight.

Volcanic activity can also cause changes to our weather and climate. When an eruption occurs large volumes of ash can be released into the atmosphere and block out the Sun's rays leading to cooler weather which can last for many months. However volcanic eruptions also release carbon dioxide and methane, which are greenhouse gases that contribute to the greenhouse effect.

Solar output has been suggested as another cause of climate change. However, although we know that the number of sunspots, which are cooler darker areas on the Sun's surface, varies over an 11-year cycle, further research is needed to determine the effect of these changes on our climate.

Although natural events such as volcanic and solar activity do have a small effect, they are unlikely to be the main factors resulting in climate change. It is now widely accepted that it is human activity and the release of carbon dioxide and other greenhouse gases into the air that is the main cause. The rise in global temperatures has been linked to the increase in the amount of greenhouse gases in the atmosphere. Before 1750, the amount of carbon dioxide in the atmosphere was around 280 parts per million (ppm). It is now over 400 ppm. There is evidence to suggest that the level of carbon dioxide in the atmosphere is higher now than at any other time in the last 2 million years.

The big change in human activity, that the Industrial Revolution started, was the burning of fossil fuels such as coal to power steam engines. Later on, oil and gas were used for transport, heating and manufacturing a wide range of goods. Burning fossil fuels produces energy, but also releases greenhouse gases into the air. Over time, large quantities of carbon dioxide have built up in the atmosphere. The carbon dioxide absorbs the heat from the Sun, warms the air and contributes to the greenhouse effect, which is when the Earth's atmosphere is warmer and moister than it should be.

Agriculture is another human factor that is causing climate change. Farming is an economic activity which uses a lot of energy to grow and produce food and so is responsible for high carbon emissions. Greenhouse gases are released when fossil fuels are used in the manufacture of fertilizers and pesticides, and in the transport of raw materials and food. In addition, cattle are responsible for emitting methane. This situation is set to continue, as the increasing world population necessitates the production of more food.

Deforestation is another cause of climate change. All over the world, trees are being cleared to make way for new roads, homes and farms. Forests store carbon and when they are burnt or felled, carbon dioxide is released into the atmosphere adding to the build-up of greenhouse gases.

The effects of climate change on people vary across the world and will depend on the relief of the area and the wealth of the country. The impacts will be larger in low lying middle-income countries such as Bangladesh than in high-income countries with areas of highland like the UK.

As the ice sheets and glaciers melt sea levels are set to rise. Low-lying coastal areas will be flooded displacing millions of people. The warmer moister air will result in more intense rainfall leading to flash floods, which is a problem in cities without open spaces and drainage. The monsoon rain can become heavier, leading to widespread flooding as was seen in Pakistan in 2022. These floods affected whole villages and extensive areas of farmland causing homelessness, food and water shortages, and the outbreak of disease.

Droughts and hot dry summers will make it more difficult to grow crops in certain areas, and yields are likely to be lower. However, growing seasons can become longer, particularly in some countries such as the UK, resulting in higher yields. Heatwaves could also become dangerous if the temperatures stay very high for long periods. Higher sea temperatures can lead to an increasing frequency of El Niño, the name given to a period of warmer seas in the eastern Pacific. The El Niño winds blow off the ocean bringing warmer than average weather particularly in December to many places.

The effects of climate change on the environment are more noticeable. As the land, air and oceans warm up, the ice sheets and glaciers will start to melt, releasing freshwater into the sea. This will cause sea levels to rise and for the sea water to become less saline or salty, which can slow or change ocean currents. As the oceans absorb more carbon dioxide, they become less alkaline, a process called ocean acidification. Ocean acidification can have negative effects on marine organisms, like coral and plankton, which are an important part of the food chain.

The management of climate change involves both mitigation and adaptation. Mitigation reduces the causes of climate change by reducing or eliminating the long-term risks to human life and property. Building earthquake-proof buildings and reducing carbon emissions are some of the actions that can be taken to mitigate climate change. Adaptation is how people respond to the long-term changes in the weather and climate. Communities can make adjustments to their lifestyles and buildings so that any damage caused by climate change is reduced and that people are able to cope with the consequences and impacts of any changes in the climate. Decisions can also be taken to take advantage of the opportunities that arise. For example, new types of crops may be able to be grown if conditions become warmer and wetter.

Alternative energy production is a key mitigation strategy. Instead of burning fossil fuels such as coal, oil and gas to generate electricity, renewable energy sources can be used instead. Nuclear, hydro, solar, wind, geothermal, and tidal power are all considered to be sustainable sources of energy as they do not emit large amounts of carbon dioxide.

Carbon capture and storage (CCS) are technologies that in future could prevent up to 90% of industrial carbon emissions from entering the atmosphere. A new CCS facility on the east coast of England will capture carbon dioxide from factories in the Humberside region and gas-powered stations on Teesside. The carbon dioxide will be converted into a liquid and stored in the empty gas fields under the North Sea.

More trees could be planted because trees act as carbon sinks, removing carbon dioxide from the atmosphere during photosynthesis. They also release moisture into the air, producing more cloud and reducing incoming solar radiation.

International agreements to reduce carbon emissions are another mitigation strategy. In 1997, 162 countries signed the Kyoto Protocol. The 2015 Paris agreement, which 194 countries signed, made a commitment to keep the rise in global temperatures to below 2°C. High-income countries pledged to help low-income nations adapt to climate change. In 2022, at the 27th United Nations Climate Change Conference (COP27), 197 countries agreed to cut their carbon emissions in order to limit the rise in global temperatures to 1.5°C by the end of the century.

Low level mitigation measures include cutting down on food waste and increasing recycling, introducing energy-saving measures at home, walking or cycling instead of taking the car, and buying local food to cut down on food miles.

Adaption to climate change is another important strategy. Steps can be taken to prepare people and places for the impacts of a changing climate. Some of the harmful changes, such as rising sea levels, can be prepared for by moving communities to safer places. Better planning and preparation can be done so that fewer lives and homes are lost when a tropical storm or flood occurs. It may also involve making the most of any opportunities, for example, longer growing seasons and increased yields in some regions.

Agricultural systems may have to change. Farmers will have to adapt as some crops may not grow well or at all in a warmer or drier climate. Drought resistant plants may have to be grown where rainfall is lower and more unreliable. Cultivation might be shifted to new areas and irrigation increased as rainfall patterns change. New crops, for example oranges and grapes, may become possible to be planted.

Water supply can be better managed. For example, water transfer schemes could be used, to move water from an area of water surplus to an area of water shortage. There can be better management of water resources to ensure evaporation and loss are kept to a minimum. Some countries are installing desalinisation plants to cope with water supply problems.

The risk from rising sea levels can be reduced by building sea defences to protect the land from being eroded away. Houses built on stilts may help to protect properties in vulnerable areas.

In conclusion, the main cause of climate change is the build up of carbon dioxide and other greenhouse gases in the atmosphere. It is likely that we will have to use mitigation to address this root cause of climate change as well as adaptation, so that the risks posed by the consequences of climatic changes are lowered. Both approaches will be necessary, because even if emissions are dramatically decreased in the next 20 to 30 years, adaptation will still be needed to deal with the global changes that have already started.

6. The living world: ecosystems

E cosystems are biological environments where life is created and sustained. In each ecosystem, living things have their own ecological niche. Plants and animals interact with non-living things, the landscape and the weather as well as each other. Some ecosystems, such as tropical rainforests, contain many diverse species of plants and animals. Other ecosystems, like those found in cold environments, have fewer species. However, in all ecosystems the communities of plants and animals that live there, depend on each other and the physical environment for survival.

Ecology is the study of the relationships between all living organisms, including humans, and their physical surroundings. The term biodiversity is used to describe the variety of life in a particular habitat or ecosystem. This includes all the living organisms such as the wildlife, plants, fungi and bacteria. Wildlife includes all the animals found there.

Ecosystems exist at a range of scales, and involve the interaction between biotic and abiotic components. They range in size from the micro to the macro. Examples of small ecosystems include ponds, hedgerows, cliffs and woodlands. Large ecosystems include tropical rainforests and deserts. Whatever the size of the ecosystem, it will have a distinct set of components which interact together and with the physical environment.

Biotic components are all the living organisms found in an ecosystem. This includes the birds, insects, animals, plants, fungi and bacteria. Every ecosystem has its own particular set of living things. Each component within the ecosystem has a specific role, whether it is consuming another organism or decomposing a dead one.

Abiotic components are the non-living physical elements or things of an ecosystem. These include temperature, amount of rainfall and light, type of soil, altitude, prevailing wind, and the amount of nutrients. These variables influence where each organism can survive and reproduce.

The interaction between biotic and abiotic components determines the nature of an ecosystem. Plants do not grow well in places which are hot and dry, but they grow rapidly where there are high temperatures combined with high rainfall. Thus, there is a wider range of biotic components within a rainforest than in a desert.

Within an ecosystem, the abiotic components can vary resulting in a different set of biotic components. For example, in a pond, plants which have adapted to growing with less light grow in the deeper parts, whereas nearer the surface there are light loving plants and creatures. The parts of an ecosystem which have different abiotic components from the rest are called niches.

The concept of interrelationships within ecosystems is that the life-cycles of plants, animals and other organisms are linked to each other as well as to the non-living parts of the environment. They form a natural system with inputs, outputs, energy flows and nutrient cycling.

The Sun provides the energy for an ecosystem to function. The Sun's rays not only warm the land but are used by the plants and trees during photosynthesis. The Sun's energy is transferred through the ecosystem via the many food chains. The energy is used by the biotic components for respiration, movement and the production of heat. All the components of the ecosystem work together to maintain the cycle of life.

Epping Forest, a deciduous woodland in south-east England is an example of a small-scale UK ecosystem. There, the older taller trees such as oak, beech and elm form the canopy layer which can be 40 metres high. Underneath is the sub-canopy, which is composed of the shorter trees such as hazel, holly and ash. Beneath the sub-canopy, is the herb or shrub layer where brambles, grasses and bluebells grow. These flower early in the year, before the trees are in full leaf and block out the light. Mosses, fungi, flowers and lichens can be found in the dark, damp ground layer.

The biotic components of Epping Forest include trees, shrubs, mammals, birds, insects and fungi. The abiotic components of the forest include a fertile brown earth soil; a temperate climate with mild winters (temperatures average 5.5°C in January), warm summers with temperatures of 18°C in July; and 690 mm of rain each year.

Within the forest are producers and consumers. A producer is an organism or plant that is able to absorb energy from the Sun through photosynthesis. In photosynthesis producers convert the Sun's energy, water and carbon dioxide into carbohydrates and glucose, a type of sugar. Producers not only produce their own food but also food and nutrients for consumers to eat. They are at the bottom of the food chain. Within Epping Forest the producers include all the trees and plants such as the oaks, brambles, ivy, lichens and bluebells.

A consumer is a biotic component in an ecosystem that gets the energy and raw materials it needs by eating producers or other consumers. A herbivore is a consumer that just eats plants. A carnivore is a consumer that eats other animals. Within Epping Forest the consumers include wood peckers, squirrels, badgers and owls. They all pass on the Sun's energy through the food chain.

A decomposer is an organism such as a bacterium or fungus that breaks down organic waste and dead tissue. A wide range of things including autumn leaves, dead animals and excreted materials are broken down by decomposers. Decomposers play a vital role in the recycling of nutrients as they convert dead matter into a form that can be re-used and absorbed by vegetation. They also release energy from plants and animals so that it can be recycled and returned to the soil. Decomposers operate at all levels in the food chain.

A food chain is the connection between a producer and a consumer. The different plants and animals that live in an ecosystem rely on one another as their source of food. An example of a simple food chain is a plant which is eaten by a caterpillar. The plant is the producer and the caterpillar the primary consumer or herbivore. A longer food chain would be if a bird, a secondary consumer, eats the caterpillar. The energy that the plant has produced from sunlight is transferred first to the caterpillar and then to the bird.

At the beginning of the food chain there are enough producers to support a large number of herbivores. However, each link of the chain uses energy and so after three or four links there is only enough energy left to support a small number of consumers at the end.

In Epping Forest a three-link food chain occurs when an oak leaf is eaten by a snail, which in turn is eaten by a thrush. This becomes a four-link food chain when the thrush is eaten by a hawk. The oak leaf is a producer; the snail a primary consumer; the thrush a secondary; and the hawk a tertiary consumer.

A food web is a complex pattern of different food chains between plants and animals in an ecosystem. A food web forms a complex hierarchy of plants and animals that rely on each other for food. There are many connections between the different biotic components within a food web. For example, the oak trees in Epping Forest are part of a complex food web. Over 2,000 biotic components including birds, fungi, invertebrates, lichens and mammals can depend on an oak for food. Squirrels and caterpillars live amongst the branches feeding on the leaves and acorns. The soil and leaf litter under the trees is home to decomposers such as earthworms and snails, which are eaten by birds, shrews, frogs and foxes. In turn the small birds and shrews are food for owls and other birds of prey.

The nutrient cycle is a set of processes whereby the nutrients in the soil are used by plants and trees to grow before being returned back to the soil. The process starts when the roots of trees grow down into the bedrock. The rocks are broken up and nutrients are released into the soil. The roots of plants and trees take up these minerals and use them to grow. The nutrients pass through the food chain when the producers are eaten. In the autumn, when the leaves fall from the trees and decompose, the earthworms mix the nutrients back into the soil. In the spring the cycle of growth starts again.

The balance between components in an ecosystem has been achieved, when all the biotic components are living sustainably without one component dominating and driving out another. When the ecosystem is in balance it is able to adapt to slow changes from human activity or climate change. However, rapid changes can upset the equilibrium.

The impact on the ecosystem of changing one component can be high. For example, if deer disappear from a forest the ash seedlings would remain un-eaten and could grow into trees. In the autumn, when ash leaf litter builds up on the woodland floor it is eaten by earthworms, increasing their number. An increase in earthworms would lead to more food for badgers resulting in an increase in the badger population.

If a stable ecosystem experiences a change in an abiotic component the impact can also be high. For example, if the oxygen levels in a lake fall, some of the plants will die, because of the lack of oxygen. The shortage of food will result in some of the primary and secondary consumers in the pond dying. The whole food web can be affected and an imbalance in the ecosystem will occur.

An overview of the distribution and characteristics of large scale natural global ecosystems illustrates the richness and variety of life on Earth. A global ecosystem is a very large ecological area or biome on the Earth's surface where the fauna and flora, which are the animals and plants, are adapted to their environment. There are several such ecosystems. They include the tropical rainforests, the hot deserts, the polar regions, the tundra, areas of deciduous and coniferous forest, the Mediterranean zones as well as the temperate and tropical grasslands.

Tropical rainforests are distributed in a broad belt of low pressure stretching from the Equator to latitudes 25°N and 25°S. They are found in Africa and south-east Asia, although the largest single area of tropical rainforest is in South America. The characteristics of tropical rainforests are high temperatures and high rainfall all year round. In these hot wet conditions, there is a continuous growing season with rapid plant growth and high levels of biodiversity.

Hot deserts are clustered along latitudes 30°N and 30°S, near the Tropics of Cancer and Capricorn, where there are belts of high pressure. There are large deserts in Africa, Asia, Australia and North America. The largest hot desert is the Sahara in Africa which spans the whole width of the continent. The characteristics of hot deserts are low rainfall – less than 250 mm of rain per year, high day time temperatures and low night time ones. Plant growth is slow in these hot dry conditions and there are low levels of biodiversity. The plants that grow in hot deserts include shrubs and cacti which are sparsely distributed in the sandy soil.

Cold environments, which include polar and tundra, are found in the high latitudes of the Earth. Polar ecosystems are located in the Arctic, Alaska, Greenland and the Antarctic where the air pressure is high. Tundra ecosystems are found mainly in the high latitudes between 60°N and 70°N, in northern Canada and northern Europe. The characteristics of polar areas are low rainfall, very low temperatures, slow plant growth and low levels of biodiversity. Six months of the year have no sunlight and the Sun remains low in the sky. The snow and ice reflect the Sun's light and heat back into the atmosphere. The characteristics of tundra are low temperatures, low rainfall and high levels of wind. The ground is permanently frozen and is called permafrost. The growing season is very short for the plants which include mosses, lichens and grasses.

Deciduous and coniferous forests are located at latitudes 50°N and 50°S where the air pressure is low. These forests are found in Europe, North America and Asia. The characteristics of these forests are medium temperatures and average rainfall. The winters are cool with average temperatures of 0.5°C and the summers warm with temperatures between 15°C to 20°C. Rainfall is around 650 mm a year. There is steady plant growth in these conditions and there are medium levels of biodiversity. In the deciduous forests the trees drop their leaves in winter so that they retain moisture.

Temperate grasslands are distributed between 30°N and 40°N, and 30°S and 40°S in inland areas. Temperate grasslands are found in North America where they are called prairies, and in Asia where they are called steppes. The characteristics are hot dry summers and cold winters. Only grasses grow in this ecosystem – the conditions are too dry for trees.

Mediterranean ecosystems are mainly found between latitudes 40°N and 45°N in places around the Mediterranean Sea and in California, South Africa and Australia. The characteristics include hot dry summers and warm wet winters. Typical Mediterranean vegetation includes olive trees and citrus trees.

Tropical grassland is also called savanna, and is distributed between 15°N and 30°N, and 15°S and 30°S. These grasslands are found in South America, Africa and Australia. The characteristics of these grasslands are high temperatures all year round, and wet and dry seasons. In the main it is too dry for most trees.

7. The living world: tropical rainforests

Tropical rainforest ecosystems have a range of distinctive characteristics. These include high temperatures, high rainfall, dense vegetation, poor soils, and high levels of biodiversity. The Sun's heat and light power the water, carbon and nutrient cycles enabling plants and trees to grow rapidly and continuously. There is a wide variety of life living in the rainforests because there are many different habitats both in the trees and on the ground.

The physical characteristics of the rainforest ecosystems include the climate, soils and vegetation. Rainforest ecosystems have tropical climates with over 2,000 mm of rain a year and no dry seasons. The average daily temperature of 28°C never falls lower than 20°C, or climbs higher than 35°C. Most of the rainfall is convectional and is often accompanied by thunderstorms. This results in a hot and humid atmosphere.

The soils found in tropical rainforests are shallow and low in fertility, but are rich in iron. This gives them a red colour. A layer of dead leaves and waste material covers the soil. Decomposers, such as fungi and bacteria, break down this organic matter and mix it into the soil. Nutrients are therefore recycled quickly.

Tropical rainforests are lush and green all year round as most of the trees are evergreen. Rainforests have a clear structure of vegetation with four distinct layers. The main canopy, which looks like a carpet of green from above, forms the highest layer. It is made up of lianas and the tallest trees, which can grow up to 75 metres high. The occasional emergent tree towers above the canopy, which is where over 50% of the forest animals and birds live.

Underneath the top layer is the under canopy, which is full of younger shorter trees. Growing below both canopies is the shrub layer, which is composed of bushes and saplings. On the cooler ground, there is a thick layer of decomposing leaves lying amidst the buttress roots. Here it is damp and gloomy because the overhead trees block out most of the sunlight and the raindrops constantly drip from the leaves above.

The interdependence of climate, water, soils, plants, animals and people within the rainforest is high. The biotic and abiotic components of the ecosystem depend on one another, and any change in one, leads to a change in the others. For example, if the climate became hotter and drier, some plants and trees would grow at a slower rate and some would not be able to grow at all. If there was less water in the rainforest, the rates of decomposition, nutrient recycling and vegetation growth would all be slower.

The shallow rainforest soils depend on the decomposers which break down the constant supply of leaf litter which falls to the ground. Humus, which is the dark, organic material that forms in soil when plant and animal matter decays, helps the soils maintain their structure and fertility. If the rainforest is deforested, the soils become dry and thin making soil erosion more likely.

The flowers, leaves, seeds and nuts of the rainforest plants and trees provide a plentiful supply of energy, shelter and nutrients for the animals all year round. If there were fewer producers, the number of consumers would fall.

The animals and birds are part of the complex food web which exists within rainforest ecosystems. They are primary and secondary consumers and they pass energy through the system. They aid the reproduction of the plants and trees by pollinating the flowers and dispersing their nuts and seeds. When they die and decompose, they supply nutrients to the rainforest.

Small groups of people live sustainable lifestyles in the rainforests, hunting and gathering, and practising shifting cultivation. They care for the forests and help to keep the ecosystem in balance. However, if the numbers of people increased and they cleared too much of the forest in order to grow more food, this could harm the ecosystem.

How do plants and animals adapt to the physical conditions found in the tropical rainforests? The plants and trees are specially adapted to the abiotic components of the ecosystem. Rainforest trees have shallow roots which can quickly absorb nutrients from the top layer of the soil. As the trees grow so high, they have wide bases and buttress roots to support them and to keep them stable. As there is a lot of rain, the trees have smooth barks and the leaves have drip tips so that the water quickly runs off them.

Birds and animals have also adapted to life in the dense rainforest. Parrots have sharp beaks to crack open nuts. Their large toes enable them to grip to the tall trees when they sway in the wind. Monkeys have strong tails to help them to swing through the trees, and flying frogs have web-like feet, which allow them to glide through the air from tree to tree. Leopards are excellent swimmers and are camouflaged to blend into the forest.

Sloths spend most of their time high up in the trees because there are fewer predators there. As well as being nocturnal, sleeping through the day to conserve energy, they also eat tough leaves which they digest very slowly. They have long clawed arms ideal for a life in the canopy but are also good swimmers if ever they drop into fast flowing rivers. Anteaters live on the dark rainforest floor. They have adapted to the low light levels by having a good sense of smell and hearing in order to detect predators.

Tropical rainforests have a very high level of biodiversity with a huge variety of life. Within just a small area there can be over one thousand species of trees and plants including herbs, orchids, lichens, teak, rubber and palm trees. The producers grow rapidly in the hot wet conditions and provide plenty of food in the form of leaves, nuts and seeds for the many forest consumers including birds like toucans, spider monkeys, gorillas, elephants, and leopards. A huge number of insects live in the rainforests including beetles, scorpions, and butterflies. Other consumers include reptiles and amphibians such as poison dart frogs, boa constrictors, iguanas, chameleons and crocodiles.

The issues relating to the biodiversity of a tropical rainforest include deforestation, fragmentation, over-exploitation, invasive species, and climate change. These all pose a threat to the rich biodiversity of the rainforests.

Deforestation, which is the chopping down and removal of trees to clear an area of forest, has a big impact on biodiversity. When the rainforest ecosystem is cleared it is lost for ever, resulting in a decline in a high number of species and the extinction of a few. The rate of deforestation around the world is declining in many places, for example in Europe and Asia. Since 2000, the rate has fallen from over 0.2% per year to 0.1%. In China the forests are expanding by 2 million hectares a year. However, in Brazil the rate of deforestation has increased in recent decades.

Fragmentation of the rainforest also can lead to a loss of biodiversity. If only isolated pockets of rainforest are left, animals find it harder to survive because there is a narrower choice of food, habitat and mating partner. Over-exploitation of the rainforest can also result in a loss of species and habitats. This is because if too many trees are felled it is harder for the forest to regenerate naturally.

Invasive species can also threaten biodiversity. If a dominant species takes over a rainforest, other species may be unable to compete for space, food or water and start to decline. This happened after the golden mussel, a type of freshwater mollusc from China, was accidently introduced to the Amazon rainforest. As the mussel had no natural predators its numbers grew forcing out the native mussels from the ecosystem.

Climate change can lead to forest degradation and a decline in biodiversity. If rainfall totals fall, tree growth will be slower and some trees will die leaving a sparser vegetation cover. Fewer trees mean less evaporation, transpiration and rainfall, eventually leading to the loss of the rainforest ecosystem. Global warming due to the increase in greenhouse gases in the atmosphere is leading to an increase in more extreme weather. These changes are likely to negatively affect the rainforest ecosystems.

The Costa Rican tropical rainforest is a case study which illustrates the causes and impacts of deforestation. The rainforest used to cover most of Costa Rica, but now it is fragmented into smaller areas, such as the Parque Nacional Tortuguero near Limón, on the coastal plain of the Caribbean coast. The causes of deforestation include subsistence and commercial farming, logging, road building, mineral extraction, energy development, settlement, and population growth. These have all played their part in reducing the size of the original rainforest.

Subsistence farming is a type of agriculture producing food and materials for the benefit only of the farmer and his family. The indigenous people of the Costa Rican rainforest have practised subsistence farming for centuries. The farmers first make a small clearing in the forest, by slashing down the trees and burning them. The ash and cinders fertilise the soil and preserve the seeds. A wide variety of different crops including coffee, chocolate and vegetables are then planted. When the soil loses its fertility, the people abandon the plot allowing it to regenerate naturally. As this is a sustainable type of farming, the rainforest remains largely unaffected. It was only when the population of Costa Rica grew that subsistence farming led to deforestation.

Commercial farming is growing crops to sell for a profit to retailers or food processing companies. In Costa Rica commercial farming has caused widespread deforestation. In the 19th century, large areas of the rainforest were cleared to make way for plantations of bananas, pineapples and melons. As the demand for fruit grew, the plantation owners increased production by clearing new areas of the forest. By the end of the 20th century, over 40% of the rainforest in Costa Rica had been cleared.

Logging is the business of cutting down trees and transporting the logs to sawmills. It can also lead to deforestation. In the Costa Rican rainforest, there are 500 species of trees, including the tropical hardwoods of mahogany and teak. Logging companies have permits to fell the hardwoods which are exported for use in the furniture and construction industries. They also own forest plantations and harvest the timber to be made into wood pellets and paper.

Road building is another cause of deforestation. Roads are needed to move goods and people around the country. When a road was built from the port of Limón to San José, the capital and largest city of Costa Rica, areas of rainforest were cleared for its construction.

Mineral extraction is the removal of solid mineral resources from the earth. The minerals which are mined or quarried include ores, stones and solid fuels. Some ores contain metals such as iron and aluminium which are commercially valuable. Precious stones, such as diamonds are also a valuable commodity. Rocks and stones such as granite are also extracted for building. Other minerals which are mined include solid fuels, such as coal and oil shale. In north-west Costa Rica there is an important gold mining industry which led to the clearance of parts of the rainforest.

Energy development, when dams, reservoirs and hydro-electric power stations are constructed, also causes deforestation. The hydro-electric power plant at Lake Arenal in the centre of Costa Rica now generates 17% of the country's electricity, but a large area of the forest was flooded when it was built.

Settlement and population growth have both been responsible for deforestation. The population of Costa Rica, which stood at 5 million in 2021, is increasing at a rate of 1.5% per year. Rainforest has had to be cleared to make way for new towns as well as the homes, schools and hospitals needed by the increased number of people.

Deforestation has enabled the economic development of Costa Rica to occur. Parts of the rainforest were cleared to allow economic activities such as commercial farming, logging and mining to start. These industries produce raw materials such as gold, timber and bananas for export, earning foreign currency, and creating jobs and opportunities. This leads to the multiplier effect, which is a cycle of economic upturn. The workers and businesses spend their money on goods and services, and pay taxes. The government uses their increased income to fund new infrastructure and public services. This economic activity has resulted in positive economic and social impacts as the Costa Rican people have enjoyed regular incomes and higher standards of living.

However, soil erosion and climate change are both negative impacts of deforestation. Soil erosion is the removal of fertile topsoil faster than it can be replaced. Soil erosion can be caused by water washing the soil away or the wind blowing it away. Animals over-grazing an area and humans farming and felling trees can also cause soil erosion.

In the rainforest, soil erosion results when the trees, which provide shade and anchor the soil, are removed. The Sun bakes the bare ground hard preventing the rain from soaking in. The water instead runs off the surface washing away the soil into streams and rivers. This causes the build-up of sediment in water courses, and contributes to flooding.

Climate change is another impact of deforestation. Tropical rainforest ecosystems play an important role in preventing further global warming. Any reduction in the amount of forest cover, results in an increase in the amount of greenhouse gasses in the atmosphere and the likelihood of more extreme weather events.

Tropical rainforests need to be managed sustainably so that uncontrolled and unchecked exploitation does not cause irreversible damage. Sustainability can be achieved by ensuring that the actions taken and forms of progress made by the present population, do not reduce the ability of future generations to meet their own needs.

The value of tropical rainforests to people is large. They are an important source of hardwoods, nuts, fruit, and rubber. These resources provide food and can be used to make medicines, homes and manufactured goods. Rainforests are also of value to the environment. The trees produce oxygen and are a vital component of the water and carbon cycles. They protect the soil, help to reduce the greenhouse effect, and slow climate change.

The strategies which can be used to manage the rainforest sustainably include selective logging, replanting, conservation, education, eco-tourism, international agreements about the use of tropical hardwoods, and debt reduction.

Selective logging is the cutting out of trees which are mature or inferior, to encourage the growth of the remaining trees in a forest or wood. It involves carefully choosing which trees to fell. Only the mature, high value trees such as mahogany are cut down and the younger trees are left to grow to maturity. This protects the ecosystem and prevents soil erosion.

Replanting takes place after all the trees in an area have been cleared for logging. Saplings are planted to replace the felled trees and are allowed to grow to maturity before being harvested. In tropical hardwood plantations, commercial farmers ensure a continuous supply of timber by planting young trees every year. Farmers, who practise agroforestry where trees and crops are gown together, also replace old trees with new ones.

Conservation is another strategy that is becoming more important. National Parks and nature reserves are created and the species living within these designated areas are given special protection. In schools and agricultural colleges, students learn about the value of biodiversity and of the environment so that they can become future guardians of the rainforests.

Eco-tourism is small-scale tourism to places rich in biodiversity such as rainforests. For example, in Costa Rica small groups of tourists visit the rainforest to see the wildlife in their natural habitats. Eco-tourism tries to minimise the impact of tourism on the environment and ecosystems. To achieve this aim, visitors stay in simple forms of accommodation such as wooden huts, which are often powered by renewable energy. The basic tenets of sustainable tourism are practiced by the eco-tour operators, for example the recycling of resources, the consumption of local food and drink, the employment of local people, and the implementation of codes of conduct for visitors and staff. There is often a focus on conservation projects which help to protect natural habitats and endangered species. Eco-tourism can have several positive environmental and economic impacts on an area. These include earning money, which can be used to manage the natural environment more effectively, and creating employment for local communities.

International agreements can also help protect the rainforest. The International Tropical Timber agreement ensures that the tropical hardwood timber on sale has come from sustainably managed legal forests and not from illegal logging. The Forestry Stewardship Council only certifies timber which has come from sustainably managed forests.

Debt reduction is when countries with large debts can have them reduced by richer countries in return for agreements to protect their rainforests from further deforestation. For example, Brazil has agreed to protect parts of the Amazon in a debt swap with the US.

All these strategies are useful ways of managing rainforests sustainably. One strategy on its own is usually not enough – there has to be a combination of several things. The economic development of the rainforests can benefit both local communities and the environment. However, the type of economic development that occurs is important. In general, small scale economic activities such as agroforestry and eco-tourism have a lower impact on the environment than large-scale commercial logging and mass tourism.

8. The living world: hot deserts

Hot deserts have the most hostile environments on Earth – because of the high average temperatures and very low precipitation that are found in these parts of the world. Hot desert ecosystems have a range of distinctive characteristics including extreme temperatures, low precipitation, high winds, low amounts of cloud cover, sparse vegetation, dry sandy soils, and low levels of biodiversity.

The physical characteristics of a hot desert include the climate, soils and vegetation. The climate in hot deserts is extreme. During the day it is very hot, with temperatures sometimes hitting 50°C. At night it is cold as temperatures can plummet to below 5°C. Deserts have a large temperature range because the atmosphere in these places contains little humidity and there are few clouds to block the Sun's rays. Deserts receive twice the amounts of solar radiation received by humid regions and lose almost twice as much heat at night.

In hot deserts, not only is rainfall very low – under 250 mm a year – but it is also unpredictable. Some years will have no rain at all but others will have all their yearly total in one short and violent downpour. The air is hot and dry. It can be windy with the wind whipping up tremendous sand storms. During the heat of the day the intensity of the sunlight is high.

Desert soils cover the terrain which can be sandy, stony or rocky. The soils are shallow and permeable – they can absorb the heavy downpours of rain, when they occur. However, the soils retain little moisture, and are salty due to the high levels of evaporation which draws salts to the surface. There are low levels of nitrogen and nutrients in the soils. This is because the scarcity of plants means that there are only small amounts of organic matter for the fungi and bacteria in the soil to decompose. Desert vegetation is sparse and spread out across the landscape. This is so that the plants and the occasional tree have enough water to survive. Typical plants include thorny shrubs and cacti.

The interdependence of climate, water, soils, plants, animals and people is high within a hot desert ecosystem. The biotic and abiotic components depend on one another, and any change in one, leads to a change in the others. For example, if the climate became cooler and wetter, some plants and trees would grow at a faster rate but others would die. If the rainfall was higher in a hot desert, the rates of decomposition, nutrient recycling and vegetation growth would all be faster.

Water is precious in a hot desert ecosystem. Surplus water is stored by plants, animals and people until it is needed. Heavy rainstorms help shape the desert by scouring deep channels in the dry landscape. Rain soaks into the underground aquifers. In the places where the groundwater comes to the surface, a fertile green area called an oasis is found.

The desert soils depend upon the few plants that there are to fix chemicals from the air into the soil. When the plants die, they decompose very slowly but they do enrich the soil with nutrients.

Plants provide nutrients, shelter and water to the animals that eat them. Desert vegetation helps the soils retain moisture, by providing shade and reducing evaporation. Their roots bind the soil together, preventing gulley erosion and the excessive leaching of nutrients during wetter periods. Desert trees such as willows, palm, and acacia break the force of the rain during thunderstorms and help prevent rapid soil erosion.

The animals and birds are part of the food web which exists within hot desert ecosystems. They are primary and secondary consumers and they pass energy through the system. They aid the reproduction of the plants by pollinating the flowers and dispersing their seeds. They also help supply nutrients to the desert soils when they die and decompose. Some desert animals are nomadic moving around in search of fresh pastures. Others are nocturnal only emerging at night when it is cooler.

Desert people survive by knowing where to find water and food to eat. They conserve as much water as they can and often live a nomadic lifestyle, always on the move searching for water. During a prolonged drought, the soil dries out and the wells run dry. If overgrazing, soil erosion and crop failure occur, the resulting famine can cause malnutrition and death.

How do plants and animals adapt to the physical conditions found in hot deserts? Desert plants have evolved three main strategies to live with the low levels of water. These ways of adapting are succulence, drought tolerance and drought avoidance. Plants which use succulence as a strategy are called succulents. Cacti, agave and elephant trees are examples of succulents. These desert plants have thick waxy stems which store water. They either have small hairy leaves or none at all. They have a small surface area in order to reduce transpiration. To conserve energy, they are slow growing. To prevent being eaten by thirsty animals some have sharp spines or are toxic. Others hide amongst rocks or use camouflage to avoid detection. They have shallow roots which can absorb water from the top layer of the soil and a small surface area. Some succulents undertake photosynthesis at night, when it is cooler and more humid. Others enter a period of idleness during a drought.

The drought tolerant plants, for example the mesquite, shed their leaves during dry periods. Their deep roots search for underground water. They can survive for long periods without rain.

The drought avoidance plants such as annuals only germinate in the autumn after a period of rainfall. They grow slowly during the winter and keep close to the ground to avoid the wind. In the spring they flower, set seed and die. This short life span ensures the survival of the species as does the long length of time that the seeds can wait for the rain to come.

The animals living in hot deserts have adapted to the heat and aridity. Lizards and snakes have small eyes, narrow heads, and very smooth scales to protect them from sandstorms. Their toes are fringed with spiny scales to help them run across the hot sand without sinking. They live in burrows in the ground, which keeps them cool during the day and protects them from the sun. Fennec foxes have fur-covered feet, heat-radiating ears and pale fur that camouflages them in the sand. When temperatures are very high, foxes and many other desert animals pant to keep cool. Cottontail rabbits are nocturnal only coming out at night when the temperature drops. They have long, thin ears from which they lose excess body heat.

Camels are well-adapted to desert life. Their broad feet allow them to walk easily on hot sand; their long eyelashes keep the sand out of their eyes; and they can close their nostrils during sandstorms. Camels can survive for over a week in the desert without eating or drinking because they live on the fat stored in their humps, and they do not sweat or urinate very often. The longer thicker hair on the tops of their bodies protects their skin from the sun. To keep cool during the day, camels have a large surface area, which maximises heat loss.

Hot deserts have low levels of biodiversity with a small number of different species, due to the harsh environment. Typical desert plants include the prickly pear cacti and date palm. A variety of animals including locusts, scorpions, gazelles and oryx live in the desert.

The issues related to biodiversity include overgrazing, population increase, economic development, and climate change. Overgrazing is grazing too many livestock, such as cattle, sheep and goats, for too long on marginal land. The result of overgrazing is that so much grass and bushes are eaten that the vegetation cannot regrow. When this happens, the bare soil dries out and is easily blown away forming dust and sand storms. The loss of the soil leads to a fall in biodiversity.

Population increase can lead to the need for more food and water. Farmers may have larger herds, start to farm marginal land and use more water for irrigation. If there is over-grazing and if groundwater is over extracted, then land degradation and soil erosion can occur, leading to a loss of habitats and species.

The economic development of hot deserts is bringing new industries such as tourism to unspoilt parts of the desert. Activities such as off-road driving, camping and making campfires can disturb wildlife and trample vegetation resulting in damage to habitats and a decline in biodiversity.

Climate change can threaten the delicate ecological balance of hot deserts. If rainfall patterns shift and droughts become more frequent and longer, water will become scarcer making it harder for plants and animals to survive.

The development of hot desert environments creates both opportunities and challenges. The opportunities include mineral extraction, which is the removal of solid mineral resources from the earth. These resources include ores, precious stones, building stones, and solid fuels. Other opportunities in hot deserts include energy, farming, and tourism. The challenges are the extreme temperatures, the water supply, and inaccessibility.

The Arabian Desert is a case study of a hot desert which illustrates both the development opportunities and the challenges. The Arabian Desert is the second largest hot desert in the world covering 2 million km². It lies mostly in Saudi Arabia but stretches from Jordan and Iraq in the north to Yemen and Oman in the south. In its centre is The Empty Quarter which is a huge sandy desert.

Mineral extraction is an important development opportunity. Gold, phosphates, bauxite and copper are mined at a number of locations throughout the Arabian Desert. At Ras Al Khair there is one of the largest integrated aluminium facilities in the world.

Energy has also led to the development of the Arabian Desert. In 1936 large oil and natural gas fields were discovered. Today Saudi Arabia, once a poor desert country, possesses about 16% of the world's petroleum reserves. Ghawar is the largest oil field in the world, holding over 11 billion cubic metres of oil, and is owned by Aramco the state-owned oil company. Oil and gas accounts for 90% of Saudi Arabia's export earnings, making it the world's largest exporter of petroleum. The oil industry brings in 87% of the government's income and accounts for 42% of GDP. Oil and gas are both fossil fuels and the energy transition towards renewable energy is already happening. The hot sun beams all day long in the Arabian Desert making it an ideal location for solar farms.

Farming is another development opportunity in the Arabian Desert. Under the desert there is a large aquifer. This water is used for irrigation – the central pivot method is frequently used by farmers. There are plentiful supplies of cheap energy as well. Using these resources, over 35,000 km² of the desert have been transformed into circular areas of fertile farmland. Saudi Arabia is now self-sufficient in dates, dairy products, eggs, poultry, fruits, vegetables and flowers, and is also a major agricultural exporter.

Tourism in hot deserts is growing in importance. Saudi Arabia is the second biggest tourist destination in the Middle East with over 16 million people visiting every year – generating an industry worth over $25 billion. Currently, most tourism in Saudi Arabia involves religious pilgrimages such as to the city of Mecca during Ramadan. The Hajj, or pilgrimage to the city, is one of the five pillars of Islam.

However, it is likely that in future many people will also visit the five UNESCO World Heritage Sites including the Al-Ahsa Oasis which has gardens and archaeology, and the natural attractions such as canyons and sand dunes. Activities such as desert rallies and races, sand dune climbing, hot-air ballooning and camel safaris are becoming more popular.

The mining, energy, farming and tourism industries all create the multiplier effect. New jobs are created and when the workers receive their wages, they spend their money on consumer goods and services. They also pay taxes to the Saudi Arabian government which can be used to improve infrastructure and public services. This in turn creates more jobs and leads to higher standards of living.

There are many challenges of developing hot desert environments. Until the twentieth century, the Arabian Desert was mostly undeveloped due to the high temperatures, vast distances and general lack of water.

The climate in the Arabian Desert is extreme with temperatures sometimes reaching 55°C. It is also very dry with less than 250 mm of rain per year. The very hot dry climate makes life in the desert and economic development very difficult. During the summer months it can be so hot that it is hard for people to work. Exposure to high temperatures for any length of time can cause sickness and even death.

However, many remote parts of the desert are far from any type of health care.

Water supply can be a problem but fortunately under the Arabian Desert is a vast aquifer of fossil water trapped after the last Ice Age. However, the over abstraction of water can lower the water table resulting in wells running dry.

Inaccessibility remains a challenge in the Arabian Desert. The desert has long been settled by nomadic people such as the Bedouin, a few of whom still travel across the desert with their camels. The more accessible places nearer the coast have become more developed but the remote inaccessible parts of the desert in the middle remain poorer. In recent times, some roads have been built to the remote areas, but this can be difficult due to the sand and heat. Inaccessible places can be reached by air but flights are long, infrequent and expensive. Until these remote areas have better infrastructure and services, development will continue to be slow. Even when development does happen, it may not be sustainable as desert ecosystems are so fragile.

Areas on the fringe of hot deserts are at risk of desertification. Desertification is the process by which land becomes drier and degraded, as a result of climate change or human activities, or both. It is a process which turns the fertile land of arid areas into dry unproductive desert. It is estimated that a third of the Earth's surface where a billion people live is at risk of desertification. The places most at risk of becoming deserts are on the fringes of hot deserts. These include the Sahel, which is a strip of land south of the Sahara Desert.

The causes of desertification include climate change, population growth, removal of fuel wood, overgrazing, over-cultivation and soil erosion. Climate change may contribute to desertification. If there are changes to the patterns of the ocean currents and winds, some places may start to experience lower rainfall and more frequent droughts. As global temperatures rise due to the greenhouse effect, more places will experience longer periods of intense heat. This will lead to the loss of vegetation cover, the soil drying out and increase the risk of soil erosion.

Population growth can lead to desertification. When human populations rise, more food is needed. In response, farmers increase the size of their herds or grow more crops. However, this can lead to the vegetation cover being removed and the soil becoming depleted. If the soil becomes thinner and drier it is more prone to erosion. The population of Mali, a country in the Sahel in west Africa, is growing at the rate of 3% a year. The rise in the number of people is putting pressure on the land leading to an increased risk of desertification.

Removal of fuel wood is another cause of desertification. If too many trees are removed for firewood and building, this can result in the loss of habitats, shade, root systems and biodiversity. In northern Nigeria many families rely on wood as fuel for cooking. When they cut down the trees, the soil can dry out and be eroded increasing the likelihood of the area becoming desert.

Overgrazing also causes desertification. When grasslands are exhausted, nomadic pastoralists move their herds in search of fresh pastures elsewhere. However, if there is a shortage of fresh grass, their herds graze marginal land instead. Marginal land has thinner drier soils which make it more difficult for the vegetation to regrow.

Over-cultivation and over-cropping by farmers can also lead to desertification. When the fertility of the soil is exhausted not only are there lower yields but there is more soil erosion and land degradation.

Soil erosion occurs more easily in the hot dry conditions found in hot deserts. After a long drought, the soils are so dry and thin, that they are easily blown or washed away. This is happening in the Sahel region, which includes Senegal, Mauritania, Mali, Niger, Chad, Sudan, and

Eritrea, where frequent droughts, overgrazing, and deforestation have resulted in soil erosion and desertification.

Strategies used to reduce the risk of desertification include water and soil management, tree planting and use of appropriate technology. Water management techniques can be used to collect and store rainwater. One way of doing this is to build earth dams across rivers and streams. These small dams, which have a core of impervious rocks and outer layers of permeable rocks, can be built cheaply using local materials. During the wet season, water collects behind the dams. This water can be used for irrigation during the dry season and to prevent desertification in times of drought.

Soil management techniques include planting native plants, such as marram grass, with extensive root systems on sand dunes. These plants help to hold the sand in place particularly if fencing is built around the area. Over time as the grasses die, they add nutrients and humus to the sand dune improving the soil, so allowing other grasses to move in.

Another soil management technique is to build rock walls, which are also called stone bunds, across steep slopes. When it rains, water and soil is trapped behind the walls. The rainwater seeps into the soil and a nutrient rich sediment builds up. This technique not only collects water but helps to prevent soil erosion.

Tree planting can also help to stabilise the sand and topsoil, and prevent the desert encroaching on farmland. Planting trees as wind breaks provides shade and reduces temperatures and evaporation rates. In Africa, it is aimed to plant a Great Green wall 8,000 km long across 22 countries. It is hoped that these trees will reverse land degradation, create new habitats for wildlife, sequester carbon and prevent desertification. Currently most of the planting has occurred in Nigeria, Senegal and Ethiopia. However, in Senegal where over 11 million trees have already been planted, doubts have been raised about their survival rates.

Appropriate technology, which is also called intermediate technology, is technology that is suited to the needs, skills, knowledge and wealth of local people in the environment in which they live. It usually combines simple ideas with cheap and readily available materials, especially for use in poorer countries, and is environmentally friendly. For example, solar cookers, which are devices that use the energy of direct sunlight to heat, cook or pasteurize food and drink, are becoming popular in many countries, including Nigeria. They use solar power and are an excellent alternative to cookers powered by firewood or fossil fuels. They are also cheap and efficient. It is hoped that their use will help to slow down the rate of desertification in many places.

9. The living world: cold environments

C old environments are found in the polar regions and the tundra. The polar regions of Earth surround the North and South Poles and are dominated by the Earth's polar ice caps. The northern ice cap rests on the Arctic Ocean and the southern on the continent of Antarctica. The tundra is the flat, treeless Arctic regions of Europe, Asia and North America, where the ground is permanently frozen. Lichen, moss, grasses and dwarf shrubs can grow here. Both the polar and tundra regions are fragile environments, which are places that are easily disturbed and difficult to restore if damaged. The plants growing in fragile areas have evolved in highly specialised ways to deal with the challenging conditions. As a result, they cannot tolerate environmental changes.

Both polar and tundra cold environments have a range of distinctive characteristics including low temperatures, low precipitation, thin soils, sparse vegetation, and low levels of biodiversity – the variety of life. The physical characteristics of cold environments include the climate, soils and vegetation. The climate is extreme, particularly in the Arctic and Antarctica and other parts of the polar regions. In the tundra, such as in northern Canada and Iceland, the climate is less extreme, with clearly defined seasons.

In the polar regions it is very cold all year round. Winter temperatures regularly plummet to -50°C. In the summer, it is rarely warmer than 0°C. These low temperatures are because in the high latitudes, the Sun is at a low angle in the sky. As a result, the Sun's energy is scattered, less concentrated and is reflected back into the atmosphere by the snow and ice. Due to the tilt of the earth, the polar regions receive no sunlight for up to six months of the year, but can have up to twenty-four hours of daylight in the summer months. In the tundra regions, temperatures rise to 10°C in the summer as the continuous daylight offsets the low midday temperatures. In the winter it can be as cold as -20°C.

In polar areas precipitation is under 100 mm a year and falls mainly as snow, but in the tundra, it is higher – around 380 mm – and mostly falls as rain. In both regions, the atmosphere is cold and dry because the air is too cold to contain much water vapour.

Soils are absent in the polar areas because the ground is permanently frozen and covered with snow and ice. However, in the tundra, enough of the top layer of permafrost thaws in the summer to allow the formation of a shallow layer of soil. These soils are low in fertility and are often waterlogged. Permafrost is permanently frozen ground, which is found in polar and tundra regions.

Vegetation is also absent in the polar regions but in the tundra, where it is warmer and wetter, mosses, lichens, small shrubs, and grasses are relatively common in the sunnier more sheltered places.

The interdependence of climate, permafrost, soils, plants, animals and people in cold environments is high. The biotic and abiotic components depend on one another, and any change in one, leads to a change in the others. For example, if the climate became warmer and wetter, the rates of decomposition, nutrient recycling and vegetation growth would all be faster. Large areas of the permafrost could thaw if the climate became warmer. The greenhouse gasses, which are trapped in the permafrost, would be released into the atmosphere, helping to accelerate global warming. The soils would become water logged and there would be more flooding.

Plant growth in cold environments is slow due to the low level of nutrients in the soil. This is caused by the slow rate of decomposition of organic material in the cold dry conditions. The flowers, leaves and seeds of the tundra plants supply energy, shelter and nutrients for the animals all year round. If there were fewer producers, the number of consumers would fall as well.

The animals and birds are part of the complex food web which exists within cold environments. They are primary and secondary consumers and they pass energy through the system. When they die and decompose, they supply nutrients to the soil.

People living in cold environments survived in the past by knowing where to find food and how to keep warm. Small groups of people lived sustainable lifestyles in the Arctic as hunters and herders. They conserved as much food as they could and often lived a nomadic lifestyle, always on the move searching for food either for themselves or their animals.

Today, only a few scientists live on Antarctica for short periods of time, but almost four million people live in the Arctic, caring for the environment and helping to keep the ecosystem in balance. However, if the numbers of people increased too much this could harm the fragile environment.

How do plants adapt to the physical conditions found in cold environments? Most vegetation grows very near to the ground in order to withstand the bitterly cold winds. For example, the tufted saxifrage, grows in a dense low clump which the harsh winds blow over. Whereas other plants, such as the bearberry, have sturdy stems to keep them stable in a strong wind.

To live through the very cold dry winters, many plants have hairy stems to keep in the heat and small leaves to reduce transpiration. As the ground is often frozen or waterlogged, some plants have shallow roots. To reproduce, several types of plants have underground runners or bulbs. Others – like the bearberry – have bright red berries which are eaten by birds and dispersed elsewhere.

As the growing season is so short, the Arctic poppy and other annuals, can germinate, flower and produce seeds all within a matter of weeks. To soak up the heat, the flowers of the poppy turn to face the Sun. To keep their roots warm they grow amongst sun-warmed stones.

The animals that have adapted to the extreme cold include polar bears, penguins and marine animals like whales, seals, and walrus. In the tundra, where conditions are less extreme, there are more sources of food. This means there is a wider variety of species including lemmings, Arctic fox, Arctic hares, wolves and reindeer. In the summer months midges and mosquitoes are abundant, attracting birds such as ptarmigans.

How do animals adapt to the physical conditions? Many, such as seals, are well-insulated with a layer of blubber under their thick fur to keep them warm. During the long cold winters, many animals like the Arctic ground squirrel and lemming hibernate in their burrows. Some insects avoid the extreme cold by living underground all the time. Another survival technique is to eat fewer calories during the winter. For example, reindeers only eat lichen during the cold dark months. Many birds migrate to warmer climes during the winter. For example, the Arctic tern flies south to Antarctica. Others have two coats of feathers to help keep them warm.

Polar bears are one of the top consumers and predators in cold places. To protect themselves against the cold and to reduce heat loss, they have thick layers of fat and fur, small ears, large bodies, and short limbs. Their coats are greasy to shed water after swimming; and are white to act as camouflage when they are hunting their prey on the snow. Their feet are large and paddle-like to enable them to roam across the ice and to swim in the sea. Their sharp claws help grip the ice and their razor-sharp teeth are ideal for catching and eating their prey.

Compared to other ecosystems, cold environments have low levels of biodiversity due to the very cold dry climate. However, biodiversity could increase if the climate became less extreme and there was a longer growing season.

The issues related to biodiversity in cold environments include climate change, over-fishing, habitat loss, and marine pollution. Climate change could lead to a loss of biodiversity. If more of the sea ice thaws in the summer, the animals which rely on the rich supply of fish and seals, would find it more difficult to reach their hunting grounds. For example, during the summer months, polar bears fatten up for the harsh winter ahead by feeding on the abundant supply of seal pups. If the ice covering the ocean melted, they would have to travel further afield in search of food, and their numbers could decline. Similarly, if the grazing lands of the reindeers were lost, the herds followed by wolfs and their other predators, would have to roam longer distances across the tundra in search of fresh pastures, leading to a decline in their numbers as well.

Overfishing has occurred both in the Arctic and Southern Oceans. In the Arctic, there has been a large fall in the numbers of Atlantic cod. In

Antarctica, large-scale commercial fishing has led to the decline not only in the numbers of fish but also in birds such as the albatross.

Marine pollution is another issue relating to biodiversity. If the Arctic and Southern Oceans are polluted with plastic, oil and chemicals this will affect the complex food chains that exist there and lead to the loss of species. Another threat comes from ships discharging ballast water into the sea. This can lead to the introduction of invasive species which out compete the native plants and fish, leading to their decline.

Habitat loss poses a big threat to biodiversity. If the pack ice disappears many species will lose their homes and sources of food. Walrus and ice seals are already finding it more difficult to find suitable places for resting, breeding and rearing their young. As the Arctic becomes more accessible to humans, increased economic activity and settlement will cause further loss of habitats.

The development of cold environments creates both opportunities and challenges. The economic opportunities that are possible include mineral extraction, energy, fishing and tourism. The challenges of developing a cold environment arise from the extreme temperatures, the inaccessibility of many places, and the difficulty of providing buildings and infrastructure. The infrastructure that needs to be built in a cold environment, in order to enable settlement and industry, includes roads, utilities, water supply and sewage. Fast broadband and mobile phone coverage are also important elements of infrastructure these days.

Alaska is a case study of a cold environment which illustrates the development opportunities and the challenges. Alaska is the largest and most northern state of the USA. The region covers 1.7 million km^2. In the east Alaska borders the Canadian state of Yukon; to the west is the Bering Strait and Russia; to the south lies the Pacific; and to the north is the Arctic Ocean. The tundra and polar environments of Alaska were among the least disturbed ecosystems in the world until relatively recently, when the opportunities for development began to be exploited. These include mineral extraction, energy, fishing and tourism.

Mineral extraction, the removal of solid mineral resources from the earth, is now an important industry in Alaska. Ores, which contain commercially valuable amounts of metals, such as copper, silver, gold and iron ore, and building stones, such as gravel; and solid fuels, such as oil and gas are all mined there. Gold is found in the Tintina gold belt and platinum is also mined in Goodnews Bay. The industry contributes $2 billion a year to the economy and has a positive multiplier effect. However, there are negative environmental impacts such as the scars on the landscape resulting from open cast mining and the legacy of toxic chemicals such as arsenic.

Energy is a very important industry in Alaska. Oil and gas were discovered in 1968 at Prudhoe Bay in the north. As the sea is frozen for most of the year the oil could not be shipped out of the region. Instead, the oil flows through the Trans Alaskan pipeline, which is nearly 3,000 kms long, to Valdez on the Pacific coast. There it is pumped on to oil tankers to continue its journey southwards.

The pipeline was built on stilts so that migrating animals could pass underneath. It is insulated so that the permafrost is not damaged by escaping heat. In order to keep the oil flowing it is warmed and moved along by pumps. The pipeline has bends and kinks in it to stop blockages and is also earthquake proof. In spite of these precautions, some pollution does occur.

The oil industry today supports 100,000 jobs and contributes $14 billion to the Alaskan economy. As oil is a fossil fuel, the energy transition towards using more renewable forms of energy means that hydro-electric and geothermal power are becoming more important sources of energy. Alaska has fast flowing rivers and steep gradients which are suitable for hydro power, which already provides 21% of Alaska's electricity. There is a hydro-electric power station at Bradley Lake and a geothermal plant at Chena Hot Springs near Fairbanks.

Fishing is another important economic activity in Alaska. A wide range of fish and shellfish, including salmon, halibut, crab, and trout are caught both at sea and in the rivers. Overall, the fishing industry, supports 78,000 jobs and contributes $6 billion a year to the Alaskan economy.

Tourism is another opportunity that is being developed in the cold environments. In Alaska, the beautiful scenery and the abundant wildlife, such as brown bears, attract over two million tourists every year, creating employment in the hotels and restaurants. Alaska is a popular cruise ship destination with day excursions to watch the whales. However, although the cruise passengers do spend money in the ports they visit, most of the profits the industry makes are retained by the cruise companies. An added problem is the air and water pollution and disturbance to wildlife breeding patterns that large cruise ships can cause.

There is a multiplier effect from all these economic activities – the workers buy goods and services which in turn creates more jobs. Tax revenues are higher and are used by the US government to provide public services and to build infrastructure.

However, in spite of the many economic opportunities that are available there are many challenges of developing cold environments such as Alaska. Until the twentieth century, Alaska was mostly undeveloped with many remote places still sparsely populated with low levels of economic activity due to the extreme temperatures, vast distances and difficulty of building. However, this is changing as modern technology and communications help to overcome the challenges posed by the cold and the difficult terrain.

Extreme temperatures in Alaska form a real barrier to development. In the winter, the weather can be awful with snow storms, strong winds and freezing temperatures as low as -20°C for lengthy periods. It is also very dark making it difficult and dangerous to work outside. Workers run the risk of frostbite which can result in the amputation of noses, fingers and toes.

Inaccessibility is another challenge. Some areas of Alaska are extremely remote, and the mountainous land makes access difficult and expensive. In winter, the only way to get to some towns is by ship, plane or along dangerous roads across the sea ice. Most people use snowmobiles. In summer, the roads to some places may be impassable because the ground is too soft. People living in inaccessible areas may be a long way from employment opportunities or services. For example, the 4,000 people who live in Utqiagvik, a remote town in north Alaska, are often cut off from the rest of the world in winter.

Building is difficult in Alaska due to the permafrost, which can be 400 metres deep in some places. Houses are often built on stilts with concrete supports. This is to prevent the homes from subsiding and collapsing, if the permafrost thaws and the ground becomes waterlogged. Buildings have steep pitched roofs, so that the snow slides off them, and are triple glazed and highly insulated, in order to prevent heat loss.

The provision of infrastructure is also difficult in cold environments such as Alaska. If roads are built on top of the permafrost they can collapse if a thaw occurs. To prevent this, roads are laid on a layer of gravel to give them more stability but even so the frequent frosts cause cracks and potholes. The permafrost makes it impossible to lay pipelines underground so they are laid on the surface and lagged to prevent them bursting in freezing weather.

Cold environments are at risk from further economic development in the 21st century. As global warming leads to the ice sheets melting, remote places will become accessible to humans. Economic activities such as mineral extraction and tourism will expand. As the demand for resources such as oil and gold grow, deposits will be found and extracted. As more people have the time and money to explore the world, tourist numbers will continue to rise and more cruise ships will sail the coast of Alaska. However, while some cold environments are suitable for economic development, other places are just too fragile.

What is the value of cold environments as wilderness areas? A wilderness is a natural environment, which has not been disturbed by people and is still in a pristine state. No human settlement or activity such as farming, mining or forestry is taking place in a wilderness.

Many cold environments are valuable because they are wildernesses with no invasive species, only native plants and animals, which are specially adapted to the harsh physical conditions. This means that cold wildernesses are important for scientific study. If scientists can understand how a polar or tundra ecosystem functions, they can transfer this knowledge to other ecosystems and develop better ways of managing them.

Why should these fragile environments be protected? Cold environments are important elements of the living world and a vital resource for mankind. If these ecosystems are developed there could be many negative impacts on the environment including pollution, the loss of habitats, the disturbance of wildlife and a decline in biodiversity. If the tundra or polar regions are damaged by building or transport, it can take a long time for the land to recover. The plants are slow growing and if damaged take many years to regrow. Many of the animals are specially adapted to the environment they live in and if their habitats disappear, they will be forced to move to other areas or to face extinction. If these fragile environments are protected then these problems will be prevented.

Strategies used to balance the needs of economic development and conservation in cold environments include the use of technology, the role of governments, international agreements and conservation groups.

Technology has been used to minimise the damage to the permafrost. To prevent heat from buildings and hot water pipes thawing the ground, these have been raised up. To minimise damage to the environment and to allow migrating herds of caribou to move freely across their territories, specially-designed raised pipelines have been built across the frozen landscapes.

National governments also have a role to play. The US government has protected many areas of Alaska since oil was first discovered there in the 1960s. The Arctic National Wildlife Refuge is a large protected area of outstanding natural beauty in the north of Alaska and is the world's last great wilderness. Every year hundreds of thousands of moose and elk migrate there to breed. The Alaskan fisheries and marine habitats are protected by the National Oceanographic and Atmospheric Administration. However, even with these protections in place there is pressure to develop these areas because they are rich in oil and other minerals.

International agreements can be signed so that wilderness areas are protected. An example of this is the Antarctic Treaty which was signed in 1959 and now protects Antarctica from excessive human activity. The treaty limits the number of tourists, controls seal hunting, sets fishing quotas and restricts the size of ships entering the ocean.

Conservation groups such as the World Wildlife Fund are working with governments, communities and businesses to protect the biodiversity of cold environments including the Arctic. Several voluntary organisations also care for orphaned and injured animals found in these extreme environments.

10. Physical landscapes in the UK

L andscapes are extensive areas of land which are regarded as being visually and physically distinct. The UK has a range of diverse landscapes which include uplands, lowlands and river systems. There are coastal, fluvial and glacial landscapes with their associated landforms such as bays and headlands, valleys and flood plains, corries and ribbon lakes.

The landscape of the United Kingdom of Great Britain and Northern Ireland is varied. The country is located mainly on two large islands, which are composed of rocks formed at different periods in the past. The oldest rocks, which are 3,000 million years old, are found in north-west Scotland, whereas the youngest rocks, which formed 60 million years ago, are located in south-east England.

The land has been shaped by the agents of erosion and deposition including flowing water, wind, sea and ice. The relief, which is a change in the height, steepness and shape of the land, is determined mainly by the geology, which is the type of rocks found under the soil.

An upland area is higher than the surrounding land and is usually between 200 metres and 500 metres above sea level. These areas of high land are often – but not always – mountainous. A mountain is an area of land which is over 600 metres above sea level. These upland areas are made up of tough, resistant igneous and metamorphic rocks, such as granite and slate, which are slower to erode.

An overview of the location of the major upland areas in the UK shows that most of the land over 200 metres is located in the north and west of the mainland. In Scotland, the upland areas include the Northwest Highlands, the Cairngorm Mountains, the Grampian Mountains and the Southern Uplands. Ben Nevis, in the Grampians, is the UK's highest peak at 1,344 metres high.

In England, the Pennines, the Lake District's Cumbrian Mountains, and Dartmoor and Exmoor in south-west England, are the main upland areas. In the Lake District, Scafell Pike at 978 metres high, is England's highest mountain.

In mid Wales are the Cambrian Mountains, whereas in southern Wales the Brecon Beacons and the Black Mountains are located. The highest mountain in Wales is found in Snowdonia in the north, where Mount Snowdon reaches 1,085 metres. In Northern Ireland, where there are the Mourne Mountains, Slieve Donard at 850 metres is the highest peak.

An overview of the location of major lowland areas in the UK, shows that the places under 200 metres above sea level are found along the coast, and in central and southern England. These lowland areas form plains or gently rolling landscapes and are usually composed of softer sedimentary rocks such as clay.

The most extensive area of lowland is in East Anglia where parts of the Fens are several metres below sea level. Holme Fen in Cambridgeshire is -7 metres below sea level and is the lowest place in Britain. Other lowland areas include The Wash in East Anglia and Lincolnshire, the Midlands, the London Basin, and the Vale of York. Most of the UK's major cities are located in these low lying areas

An overview of the location of the major river systems in the UK, shows that both the two longest rivers with their drainage basins and watersheds are in southern England. The Severn is the longest river, flowing 354 km, from its source in Wales and to its estuary in the Bristol Channel. The second longest river is the Thames, which is 346 km long, and which flows east from its source near Oxford to its estuary on the east coast of England.

Other major rivers include the Trent and Mersey, which both drain large areas of central England. The longest river in Scotland is the Tweed, which is 155 km long; whereas the longest river in Northern Ireland is the Bann which has a total length of 122 km. Other important rivers include the Tyne, Great Ouse, Wye, Tay, Spey, and Avon. Most of the rivers have their source in the mountain regions and flow to the sea.

The landscape of the UK has been shaped by the erosive force of these rivers. River valleys are attractive places for settlements and most of the UK's major cities are located in one, including London in the Thames valley and Liverpool on the Mersey.

11. Coastal landscapes in the UK

T he coastline is a dynamic place – it is always changing in some way. The coast, which is the land bordering the sea, is shaped by a number of physical processes including weathering, mass movement, erosion, longshore drift and deposition.

Waves, which are ripples caused by the transfer of energy from the wind to the sea, are agents of coastal erosion. The largest waves form when a very strong wind crosses a large expanse of ocean and blows for a long time. On a calm day there may be very small waves or none at all. There are two types of waves – constructive and destructive.

The degree to which waves erode the land depends on the coast's geology and relief – its height, shape and steepness – and the fetch. The fetch is the distance the wind has blown over the sea. The longer the fetch, the more powerful the winds and the higher the waves.

When the waves reach the shore, their crests rise, surge forward and break onto the beach. The seawater that carries on flowing up the beach at an angle is called the swash. The water that flows back down to the sea is the backwash. The backwash returns sediment to the sea and flows at right angles to the beach.

What are the characteristics of constructive waves? Constructive waves have long lengths, low heights, gently sloping fronts, low energy, a strong swash and widely spaced crests. When they break onto the beach they spill forward and deposit sand and pebbles, making the beach larger.

What are the characteristics of destructive waves? Destructive waves have short lengths, tall heights, high energy, steep fronts, a powerful backwash and closely spaced crests. When they crash and plunge onto the coast during a storm, they remove large amounts of sand and pebbles making the beach smaller.

Coastal processes include mechanical and chemical weathering. The coastline is being weathered all the time by the waves, wind and rain. Mechanical weathering causes the physical disintegration or break up of exposed rock, without any change in the rock's chemical composition. Mechanical weathering occurs when the sea or rain becomes trapped in the cracks of a cliff. At night the water freezes but thaws during the day. This freezing and thawing weakens the cliff. The cracks become larger and bits of rock break off and fall to the ground as scree. When salt builds up in the cracks it expands and puts pressure on the rocks, causing flakes of rock to break off. Chemical weathering is the decomposition or rotting of rock along the coast by sea water, which is slightly acidic. The sea water dissolves and wears down alkaline rocks such as chalk and limestone.

Mass movement, which includes sliding, slumping and rock falls, is a coastal process where a large amount of weathered material moves en masse. Mass movement occurs after a period of heavy rain. The water seeps into the vertical bedding planes and the lines of weakness in a cliff face. The rocks become saturated, lubricated, heavier, loose and unstable. Eventually the rocks become so heavy, they start to move down the cliff face under the force of gravity. The speed of this movement can vary considerably – sometimes it is slow; at other times it is fast. Sliding occurs when the saturated rocks flow down the cliff in a straight line. Rock falls happen when large chunks of rock break off and fall onto the shore below. Slumping occurs when a whole segment of the cliff breaks off along a shear-plane and moves downwards in a rotating movement.

Erosion is a major coastal process. There are three main types of coastal erosion. These are hydraulic power, abrasion and attrition. Each type of erosion uses a moving force to wear away the coast and remove rocks.

Hydraulic power occurs when breaking waves trap and compress pockets of air in the cracks in the cliffs along the coastline. The compressed air exerts force and pressure on the rocks and weakens them. The cracks become larger and in time the rocks break up and fall into the sea.

Abrasion is the wearing away of cliffs by the sediment flung at them by the breaking waves. The sand and pebbles in the waves batter the rocks and smooth away any roughness in them. This process is similar to rubbing sandpaper against an uneven surface.

Attrition is erosion caused when rocks and boulders, transported by waves, smash into each other, and break up into smaller, smoother and rounder pieces.

Both transportation and longshore drift are important coastal processes. Transportation is the movement of eroded material such as boulders, sand and pebbles by the sea. The sea moves the sediment along the coast and out into deeper water. Longshore drift is the zigzag movement of beach materials along the coast. This zig-zag movement occurs when sand and pebbles are driven up the beach obliquely by the swash. Some of the material is drawn back down again, at right-angles to the shoreline, by the backwash. Heavy boulders are moved shorter distances than lighter sediment, such as sand.

Deposition is the laying down of new material on the land. Deposition occurs when material being transported by the waves is dropped because the waves lose the energy to carry it any further. This loss of energy occurs when the wind drops, the waves become lower and weaker, and the swash becomes stronger than the backwash. New landforms such as beaches, sand dunes, spits and bars are created from deposition.

Distinctive coastal landforms are the result of rock type, structure and physical processes. There are two types of rock – hard and soft. Hard rocks such as basalt are more resistant to erosion, take longer to be eroded, and retain their shape and structure, for example as steep cliffs, for longer than softer rocks. Soft rocks such as clay, sandstone and mudstone are less resistant to erosion, are eroded at a faster rate and are washed away by the waves, leaving behind landforms such as bays and beaches.

The structure of the coast also plays a part in determining the type of landforms that will be found. If a coast has a discordant structure, where bands of hard rock and soft rock alternate with each other, there are more likely to be bays and headlands. However, if a coast is concordant, it will have the same type of rock along its length and fewer bays and headlands.

Physical processes also determine the type of landforms found along a coast. If the main process is erosion, then cliffs, wave cut platforms, caves, stacks and arches form. However, if the main process is deposition, beaches and spits are more common.

How do geological structure and rock type influence coastal landforms? Geological structure is a result of ancient tectonic activity. If the rocks have been folded and faulted during an earthquake, they will have cracks and lines of weakness which can be more easily eroded by the waves.

Rock type also plays an important part in determining the type of coastal landforms that are found along a coast. Harder rocks, for example chalk and limestone, erode more slowly and usually form cliffs. Weaker rocks such as sands and clays have less structure and are more easily eroded, producing a lower cliff profile with frequent mudslides and slumping.

Which coastal landforms form as the result of coastal erosion and what are their characteristics and sequence of formation? The coastal landforms which result from erosion include headlands and bays; cliffs and wave cut platforms; caves, arches and stacks.

Headlands are areas of highland which jut out into the sea and have steep cliffs on all three sides. Bays are wide curved inlets along the coast. The formation of headlands and bays occurs along discordant coastlines, where the rocks are at right angles to the coast, and where bands of harder and softer rock alternate with each other. The sea erodes the weaker rock at a faster rate than the more resistant rock. Over time the differences between the rocks becomes more pronounced, with the harder rock forming headlands and the softer rock forming bays. Beaches form in the bays because as the waves approach the shore and the shallower water, they slow down, loose energy and deposit their load of sand and shingle. Over time, discordant coastlines become straighter as the headlands are eroded away and the bays are filled with deposition.

Cliffs are steep high rock faces which have been formed by weathering and coastal erosion. Cliffs form along coasts where the rocks are harder and slower to be eroded. Harder rocks which form cliffs include sedimentary rocks such as sandstone, limestone, and chalk, and igneous rocks such as granite and basalt. When the waves hit the cliffs, a combination of hydraulic power, abrasion and attrition erodes them. This erosion is particularly fast during a storm when the destructive waves have plenty of energy. If a wave cut notch forms at the base of the cliff, eventually the cliff above becomes unstable, collapses, and retreats. The backwash of the waves carries away the eroded material, leaving a wave-cut platform behind. The process is repeated and the cliff continues to retreat.

Wave cut platforms are gently sloping rocky ledges found at the base of cliffs and headlands. These erosional landforms have many rock pools and are covered by the sea at high tide. Wave cut platforms form as the result of cliff retreat and wave abrasion. Over time, the potholes become larger and join together, lowering the platform.

Caves are round large holes usually found at the base of cliffs. Arches are wave-eroded passages of curved rocks which span an opening in a small headland. Stacks are small rocky columns standing on their own just off the shore and surrounded by sea.

Caves, arches, and stacks form in a sequence which starts when the waves attack the base of a cliff. The waves use a combination of hydraulic power, attrition and abrasion to erode any areas of weakness in the cliffs, such as joints and cracks. With continual wave attack these cracks and joints get larger and develop into caves. Sometimes two caves form on both sides of a headland. When the rocks separating the caves are eroded away an arch is formed. Over time, wave attack and mechanical weathering leads to the roof of the arch collapsing leaving a stack behind. Further erosion reduces the stack to a smaller, lower stump.

Which coastal landforms form as the result of coastal deposition and what are their characteristics and sequence of formation? The coastal landforms which result from deposition include beaches, sand dunes, spits and bars. Beaches are narrow, gently sloping strips of sand or shingle found at the head of a bay. They form in the zone of deposited material that extends from the low water line to the limit of storm waves. A beach is divided into the foreshore and the backshore. The section of a beach which is exposed at low tide is called the foreshore. The backshore is the area of the beach closest to the land. Usually waves only reach the backshore during a storm or a very high tide.

Beaches are formed by constructive waves, which are low waves with a powerful swash and a weak backwash. When these waves lose energy, they deposit sand, pebbles and shingle onto the beach. The finer sand is deposited nearer the low water mark and the bigger pebbles further up the beach. Beaches are constantly changing as tides and the weather can alter every day, bringing new materials and taking away others. Sandy beaches are usually found in bays where the water is shallower and the waves have less energy. Pebble beaches often form where cliffs are being eroded, and where there are higher energy waves.

Sand dunes are mounds of sand with a steep windward side, which usually faces the sea, and a gentle slipface on the leeward side. These sandy hills are found above the high tide mark, are shaped by wind action, and can be covered with grasses and shrubs.

Sand dunes form when wind blows sand along a beach or into a sheltered area. Grains of sand accumulate behind an obstacle like a bush or a boulder and grow larger over time. A large tidal range helps in the development of dunes – as does an onshore wind.

Spits are long narrow sand or shingle beaches that extend from the mainland out to sea. Spits often have a number of different recurved ends or hooks, each marking a former position of the spit. Behind a spit, there is often an area of very sheltered sea where sediment is deposited and the water becomes so shallow that salt marshes or mud flats form.

Spits form along discordant coasts where there are bays and headlands and river estuaries. The coastal processes which are responsible for the formation of spits are deposition and transportation. In the places where spits form the prevailing wind blows at right angles to the coastline and the waves and longshore drift follow the same direction as the prevailing wind. Deposition occurs on the leeward side of a headland where the sea is very sheltered and where sediment deposited by a river builds up in an estuary. The spit forms an area of new land which continues to grow as more sediment is added. The end of the spit changes position if there is a strong sea current or a change in wind direction.

Bars are long narrow areas of sand or shingle which are joined to the land at both ends. Bars form where a spit grows between two headlands across a bay or river estuary. This happens where there is too much deposition to be removed by longshore drift and where the strength of the waves or the river currents are too weak to transport the sediment out to sea or along the coast. The water behind the bar becomes trapped and forms a lagoon, which is a shallow lake. Lagoons do not last forever – eventually they fill up with sediment and become new land.

An example of a section of coastline in the UK to identify its major landforms of erosion and deposition is a seven-mile stretch of coast between Borth and Aberystwyth in west Wales. The geology of this coastline consists of sedimentary sandstones and mudstones, which were laid down on the sea bed 420 million years ago, then uplifted and tilted, forming new land with vertical bedding planes.

This coastline faces the Atlantic Ocean, over which south-westerly winds blow for up to 8,000 km without interruption. This long fetch results in the formation of large powerful waves, especially in the winter, when strong winds blow continuously. When the destructive waves hit the coast, their strong backwash removes sand and rocks from the land.

The coast from Borth to Aberystwyth is discordant with headlands and bays. The headlands are formed from the harder sandstones which are more resistant to erosion and the bays from the softer mudstones. Just south of Borth is Carreg Mulfran a 65-metre-high headland. Two kilometres north of Aberystwyth is Clarach Bay a small sandy bay with a sand and shingle beach 40 metres wide.

Along several stretches of the coast are erosional landforms. Steep cliffs, 100 metres high in some places, face the sea. Cut into the cliffs, are several caves an example of which is found near Clarach Bay. This cave, which is 12 m² in size, was formed in a line of weakness along a fault line, where the softer rock has been eroded. The roof of the cave is made from harder rock. At the back of the cave is a smooth-sided tunnel, formed when the sea comes in and swirls large boulders around, abrading the sides. Eventually the roof will collapse and form an arch. At Clarach Bay, landslides have occurred where the soft mudstone cliffs have slumped downwards under the force of gravity.

Underneath the cliffs is a 10-metre-wide wave cut platform. This narrow gently sloping platform can be seen at low tide. It has many large potholes some of which are over 1 metre deep. At Craigfulfran, north of Aberystwyth, there are several stacks, which are small rocky isolated landforms formed when the roof of an arch collapses, leaving behind an abandoned rock.

This stretch of coast also has several depositional landforms. There are sandy beaches at Aberystwyth and Borth. North of Borth is an extensive area of sand dunes called Borth Sands. The older sand dunes are found on the sheltered leeward side and the newer ones on the windward side facing the sea.

At Ynyslas near Borth, there is a 5 km long spit which connects the beach to the estuary of the River Dovey. The spit, which is 6 metres high in places, is growing northwards at the rate of 2 km every 100 years. The spit has a series of recurved ends and it gets narrower towards the north. The northern end of the spit is covered with sand dunes which are continuously forming. There is a reclaimed salt marsh behind the spit.

Aberdovey Bar is being formed in the River Dovey estuary. It is an offshore bar as it is further out to sea. It has been formed by the constructive waves depositing sediment due to friction with the sea bed. The Dovey estuary was once a glacial valley and it is likely that terminal moraine was deposited in the estuary enabling a bar to form. It is unlikely that a complete bar will ever form across the estuary as the tidal scour is too great.

Different management strategies can be used to protect coastlines from the effects of physical processes such as erosion. The management strategies include hard engineering, soft engineering, and managed retreat. Hard engineering strategies involve building structures which can prevent the sea from damaging the coast, particularly during storms when the waves are at their highest and most powerful. Soft engineering strategies manage coastal erosion by working with natural processes to help restore and improve beaches and coastal ecosystems. Managed retreat is a strategy which allows the coast to evolve naturally with no interference from people.

How is hard engineering used to protect coastlines? Hard engineering uses concrete and large artificial structures to defend land against coastal erosion. Civil engineers build groynes, gabions, sea walls and rock armour in order to protect a section of coast from being eroded. These structures prevent the sea eroding beaches, homes and farm land.

Groynes are long narrow wooden barriers which are built from the beach out into the sea. They are erected in order to stop the longshore drift of sand and shingle. Groynes not only keep the sand on the beach but they also protect cliffs from erosion and provide an important tourist amenity.

Gabions are steel wire cages which are filled with boulders or pebbles and placed along the bottom of a cliff to provide support. They can absorb wave energy so can act as a buffer against the sea. Sea walls are built of concrete, steel or stone and are built along sections of the coast which are being eroded. They provide a barrier which reflects wave energy.

Rock armour are large blocks of natural rock such as granite which are piled up at the base of a cliff in order to absorb wave energy and defend the coast from wave attack.

What are the costs of using hard engineering? The type of hard engineering used by the Environment Agency, land owners and local councils depends on a number of different economic and environmental factors. Each of the strategies has its own costs and benefits.

Groynes cost up to £100,000 each. This puts them in the medium category for capital costs. They also have medium maintenance costs - recent repairs at Dawlish Warren to just three of the existing 17 groynes cost £380,000. Another added cost is that their installation can speed up erosion further down the coast because they trap sediment and deprive another area, down-drift, of new beach material.

Gabions come in all shapes and sizes and can be filled with different types of materials. Installing gabions can be a cheaper option as they cost between £50 and £500 per metre. As well as these medium capital costs, their maintenance is also in the medium range.

Sea walls are expensive with high capital costs – typically £700 to £5,400 per metre. For example, the sea wall at West Kirby in the Wirral cost over £15 million. However, they have low maintenance costs especially if they are well built.

Rock armour has high capital costs – £1,350 to £6,000 a metre – but cost little to maintain. A disadvantage though is that they are often a different type of rock from the local geology and so look out of place.

What are the benefits of using hard engineering? Each type of hard engineering has its own advantages. Groynes are effective at preventing longshore drift from moving sand and pebbles along the beach because beach material piles up against the groyne on the updrift side. They protect the beach during a storm, slow down the waves, act as windbreaks and increase the size of the beach. Resorts that have built groynes report an increase in visitor numbers because families are attracted by the sandier beaches, safer swimming conditions and the possibilities for fishing. Groynes have a life span of 40 years.

Gabions can merge into the landscape when they are covered in vegetation. They absorb wave energy and can also improve the drainage of cliffs. Gabions are ideal as a quick-fix solution and they can last 20 to 25 years.

Sea walls protect cliffs from erosion and help to prevent flooding. They deflect wave energy back to sea. Once in place, they give people a sense of security and often have a promenade on top, which has an amenity value for tourists and cyclists. The steps at the base of a sea wall can act as seating areas for beach users. If well maintained, sea walls can last for many years and be very effective.

Rock armour absorbs wave energy by allowing the water to flow through them. Rock armour are quick to build, easy to maintain, have a long life and are versatile. The rocks can either be placed in front of a sea wall to lengthen its lifespan or used to stabilise slopes on sand dunes. Although rock armour can have a short life span, they are effective at preventing erosion. Rock armour can provide habitats for wildlife and blend in well with the local landscape.

Hard engineering solutions can be effective strategies as they do prevent coastal erosion when used appropriately. However, although their construction reduces erosion in one place, it may cause more erosion further down the coast. These unintended consequences can be problematic particularly where there is no coastal protection. The problem is therefore shifted, not solved.

How is soft engineering used to protect coastlines? Soft engineering includes beach nourishment, beach reprofiling, and dune regeneration. Beach nourishment is replacing the sand and shingle that has been eroded by the waves. Large amounts of new beach material are brought in from another part of the coast and added to an eroded beach.

Beach reprofiling involves using bulldozers to shift surplus sand, shingle or boulders from the lower part of a beach to the upper part or to where the beach sediment has been eroded. Storm ridges are removed, hollows filled in and the beach levelled. This is usually done in the spring after the winter storms. Beach reprofiling is carried out in order to make the beach safer for holiday makers and to protect the beach from further erosion.

Dune regeneration is the creation of new sand dunes or the restoration of existing ones. This can be done by planting marram grass which traps and stabilises blown sand, and enables the dune to grow larger. Dunes can be thatched with branches, dead trees or other low-cost materials, which stabilises the sand, reduces trampling and protects vegetation. Fences can also be built along the seaward facing sides of dunes to encourage the deposition of wind-blown sand. Boardwalks across the dunes can protect the vegetation as well. Dune regeneration is undertaken in order to strengthen the dunes and prevent excessive coastal retreat.

What are the costs of using soft engineering? The cost of these types of soft engineering varies. Beach nourishment is more expensive with medium costs to both build and maintain – typically £350 to £6,450 per metre. Beach reprofiling is cheaper with low building costs – £15 to £120 a metre – as well as low maintenance costs. However, the disruption caused by reprofiling may have negative impacts on beach habitats.

The cheapest option is dune regeneration with low-cost options of grass planting, fence building and dune thatching. However, dunes can be easily damaged by storms and it can be time-consuming planting the grass. Fencing off the dunes can also deter tourists

What are the benefits of using soft engineering? The benefits of using beach nourishment as a soft engineering strategy are that the appearance of the beach is improved and there is more sand for visitors to use. The enlarged beach helps to prevent coastal flooding as it provides a buffer zone between the land and the sea. This forces the waves to break further offshore.

The benefits of beach reprofiling are that afterwards the beach looks better and is safer for children. The benefits of dune regeneration are that sand dunes act as a physical barrier between the sea and the land. They absorb wave energy and water and in doing so protect the land from the sea.

All these soft engineering strategies can produce an attractive amenity for tourists. They may also increase biodiversity by providing a greater range of natural habitats for plants, animals and birds. Soft engineering is a sustainable solution as it does not interfere with coastal processes elsewhere or affect other areas.

How is managed retreat used to protect coastlines? Managed retreat is also called coastal realignment or the just-do-nothing option. In this strategy, parts of the coast, which are not used for tourism, fishing, settlement or farming are allowed to evolve naturally. Coastal processes such as erosion and deposition are permitted to shape the coast without any interference from people. Managed retreat may involve setting back or realigning the shoreline and allowing the sea to flood areas that were previously protected by embankments and seawalls.

What are the costs of managed retreat? Managed retreat is a low-cost option. However, unmanaged coastal processes may result in more coastal flooding and the loss of farmland and homes. Landowners and communities may need to be compensated for any economic losses they suffer.

What are the benefits of managed retreat? Where coastal deposition occurs new beaches, spits, bars and sand dunes are created. These new landforms and ecosystems can become a haven for wildlife and act as a form of coastal protection.

Minehead is a coastal resort in west England. Its tidal range of 14 metres is the second largest in the world. This UK town has implemented a coastal management scheme. This case study examines the reasons why the management scheme was needed, the management strategies which were adopted, and the resulting effects and conflicts.

The reasons the resort needed a coastal management scheme was that high tides regularly flooded the homes and businesses along the sea front during winter storms. Clearing up after the floods and repairing the storm damage was estimated to be costing the town over £1 million a year. The beach was also being eroded and there was a fear that if the beach was lost, visitor numbers would decrease affecting the town's income from tourism.

The management strategies adopted by the Environment Agency included both hard and soft engineering. A high sea wall with rock armour at its base was built along the front. This wall had a curved front to deflect the waves, and a curved top to help prevent flooding. Faced in local sandstone, it is an attractive feature with a wide walkway and seating areas alongside it. Rock groynes were built to help retain the beach and stop longshore drift moving sand to the east. Beach nourishment was used to build up the beach by 2 metres in height. The scheme was completed in 2001, at a cost of £12 million.

The resulting effects were positive – both economically and environmentally. The coastal defences now protect the town from storms and high tides. Tourism continues to be a major economic activity as the enhanced seafront and better beach attract visitors.

However, there were conflicts as a result of the scheme. Conservationists opposed the scheme because they wanted nature to take its course and for there to be no human interference with natural processes. They considered that the cost of protecting the town was too high.

Local residents and businesses disagreed with the conservationists because they wanted their homes and shops to be protected from coastal erosion and flooding. Previously they had had to move out after the damage caused by the floods was being repaired. There had been job losses when storm damage caused the local businesses to close temporarily.

While the local councils and the Environment Agency agreed that some sort of coastal management scheme was needed, there was some conflict over the cost involved and who would pay for it. In the end, £7 million came from the Environment Agency, and another £6 million from the local councils and Butlins, a major tourism company located in the town.

Overall, the coastal management scheme in Minehead was judged by many to have been a success.

12. River landscapes in the UK

River landscapes are important geographical features in the UK. River valleys are low stretches of land between hills. A river valley will usually but not always contain a river which is flowing within its channel. River valleys and channels vary in size and shape – their width and depth depend on the volume of water, the gradient and the geology of the landscape. River landscapes are dynamic – eroding the land in some places and depositing sediment in others. Rivers not only help shape the physical environment but influence the ecosystems and human activities in the surrounding areas.

Rivers start their journey to the sea at their source, which is usually a spring. As a river flows downhill it takes a path that is called its course. A river's course is split into three stages – the upper or youthful stage, the middle or mature stage, and the lower or old age stage where the river ends.

The shape of a river valley changes as the river flows downstream. Although every river and its valley are different, the shape of the valley changes in the same general way. In the upper valley, the river is in its youthful stage with a narrow channel and floodplain. The water is generally quite shallow and is flowing slowly. As the river has just begun to erode its valley, there will be large boulders and rocks in the channel.

When the river enters its middle stage, it becomes wider, deeper and faster because it contains more water and has more erosive power. In its last stage in old age, the river reaches its full force because it contains more water and is flowing at its fastest speed. The river channel is also at its widest and deepest in the lower course.

What is the long profile of a river and its valley and how does it change? The long profile is the slope that the river goes down from its source to its mouth. The angle of the slope is known as the gradient. In the upper course the valley profile can be very steep with a high gradient, but by the time the river reaches its lower course the valley floor is almost flat and the gradient is low and gentle.

What is the cross profile of a river and its valley and how does it change? The cross profile is the side-to-side cross-section of a river channel and its valley. The valley's cross profile changes as the river flows downstream. In the upper course, the valley is V-shaped with steep sides and a narrow floor, and the river channel is narrow and shallow. By the time the river reaches its middle course its valley will have become wider and lower with more gentle sides and a floodplain. The river channel will also be wider and deeper with raised levées, also called embankments, on both sides. In the lower course, the river channel will reach its widest and deepest state. The valley will also be at its widest and flattest, with gently sloping sides known as bluffs on either side.

What are fluvial processes? Fluvial processes are the actions such as erosion, transport and deposition taken by a river to shape its channel and valley. Erosion is the most powerful fluvial process especially after periods of heavy rainfall and during floods. The types of erosion include hydraulic action, abrasion, attrition, solution, as well as vertical and lateral erosion.

Hydraulic action occurs when the force of the river water traps air in the cracks and crevices in the river banks. The air pressure weakens the banks and gradually wears the sides away. Hydraulic action also breaks up rocks into smaller pieces when the sheer weight of the river flows over them.

Abrasion takes place when the rocks, pebbles and mud, that the river is carrying, rub and wear down the river bed and bank and smooth away any roughness. This process is similar to rubbing sandpaper against an uneven surface.

Attrition also happens as the river moves rocks downstream. As the rocks and pebbles flow along, they smash into each other, break up and become smaller and rounder.

Solution is a process which occurs when the river flows over sedimentary rocks such as chalk and limestone. The acid in the water dissolves the soluble particles of rocks which then are held in solution in the river.

Vertical erosion is the downward cutting of a river channel. The river erodes the rocks on the river bed and deepens the river channel. Lateral erosion is the sideways erosion by a river on the outside of a meander. Lateral erosion eventually leads to the widening of both the river and the valley, and contributes to the formation of the flood plain.

Fluvial processes also include transportation, which is a process involving the movement of eroded material such as soil, pebbles or boulders from one place to another. There are four types of transportation – traction, saltation, suspension and solution. Traction occurs where the water rolls or pushes the boulders along the river bed. Saltation is when particles of rock bounce along the bottom of the river bed using the force of the water. Suspension is very small particles of silt and clay and other fine solid material held in the water while the water is moving. Solution occurs when rocks, like chalk, dissolve into the water and are carried along in solution.

Deposition is a fluvial process which lays down sediment on the river bed and banks. Why do rivers deposit sediment? Rivers deposit sediment such as silt, sand and clay when they no longer have enough energy to carry it further downstream. This loss of energy occurs when the flow becomes too slow or the river becomes too shallow. This can happen anywhere along the river's course but is usually on the inside bend of a meander, or at the mouth of the river.

Which fluvial landforms form as the result of erosion and what are their characteristics and sequence of formation? When erosion is the dominant process in a section of the river valley the fluvial landforms which form include interlocking spurs, waterfalls, and gorges. Interlocking spurs are ridges of land which jut out on the opposite sides of a steep sided youthful valley. The characteristics of interlocking spurs are that they have steep gradients, convex slopes and are made of harder rock. Interlocking spurs form in the upper course of a river because the river lacks the erosive power to remove them. The river just weaves its way around them. Further down the valley the interlocking spurs have been eroded away by the more powerful flow of the river.

Waterfalls are streams and rivers that flow over a precipice, which is a tall and very steep rock face, down to a lower level of the river course. The characteristics of a waterfall are that they have almost vertical slopes, over which the river or stream flows downwards into a deep plunge pool at the base. The two fluvial processes which are responsible for the formation of waterfalls are hydraulic action and abrasion. Waterfalls form in places where a band of more resistant rock meets a section of softer rock. As the harder rock is eroded more slowly than the softer rock, a step in the river's long profile forms. When the river flows over this step, the water plummets to the bottom of the waterfall and erodes a large basin. Inside this plunge pool the pebbles swirl around abrading and eroding the sides.

Gorges are rocky valleys containing fast flowing rivers. The characteristics of gorges are that they are valleys which are steep sided, narrow, and deep. They are often found downstream of a waterfall and form when the harder rocks at the top of the waterfall are undercut by hydraulic action. This erosion causes the overhanging cap rock to break off. The waterfall retreats upstream leaving behind a steep sided gorge.

All these landforms are found in the upper river valley where the river is at its shallowest and the gradient is at its steepest. When water levels in the river are low, the river uses its energy to flow downstream and to transport sediment. However, after periods of heavy rain, when water levels are high, the river has enough energy to start eroding its valley.

Which fluvial landforms form as the result of erosion and deposition and what are their characteristics and sequence of formation? Meanders and ox-bow lakes are both landforms which form as a result of erosion and deposition. Meanders are pronounced wide bends or curves in a river. Their characteristics are a steeper outer bank known as a river cliff, and a gently sloping inner bank, also called a slip-off slope. Meanders form in the middle course of the river where the gradient of the valley is gentler than before. At this stage, there is more lateral erosion than vertical taking place, and also more sediment being transported.

The fastest flow of water in a river is called the thalweg. The thalweg is usually found in the middle of the river channel where it is deepest and there is less friction. Nearer the river banks, the water is shallower and slower because there is more friction.

However, when a river flows around a meander, the thalweg changes to near the outer bank. This fast-flowing water has plenty of energy for lateral erosion which it uses to undercut the river cliff, making it steeper and higher. On the inside of the bend, where the river flow is slower, material is deposited on the gentle slip off slope. The larger rocks are deposited first because they are heavier to carry. The lighter finer sediment is carried in suspension further down the valley and is deposited on the flood plain and at the river mouth. Over time, because of erosion and deposition, meanders gradually change shape, move across the floodplain, and migrate downstream.

Ox-bow lakes are former meanders that were cut off from a meandering river. The characteristics of oxbow lakes are that they are small, shallow, arc or horseshoe shaped lakes. Their formation follows a sequence. The river first develops a meander which has a very pronounced loop. The neck of this meander is narrowed by lateral erosion. During a flood, fast flowing water cuts through the neck separating the meander from the main river channel. The river flows through the neck and follows this new shorter steeper course. The cut-off meander forms an oxbow lake which gradually silts up to form an area of marshland on the floodplain.

Which fluvial landforms form as the result of deposition and what are their characteristics and sequence of formation? Landforms which result from deposition include levées, flood plains and estuaries. These are found in the lower course of the river valley.

Levées or embankments are ridges of sediment found parallel to the river and on both sides. The characteristics of levées are that they are long narrow banks with gently sloping sides. Levées form during and after a flood when the river slows down, loses its energy and deposits its load. The larger pebbles and rocks are dumped closest to the river banks forming large raised mounds. The finer sand and mud are laid down further away forming the gently sloping sides of the levées. Over time, levées get higher and wider but they can be broken through during a flood when a river reaches its bank-full capacity. Although levées form naturally when rivers flood, they may also be built up by people to protect the area against flooding.

Flood plains are the areas of a river valley which become covered with water during a flood. The characteristics of flood plains are that they are wide, flat, fertile areas of land found on both sides of a river. Flood plains form after the interlocking spurs have been removed by the lateral erosion of the meanders, as they move downstream. Deposition occurs when the river floods. After a period of heavy rain, the water level in the river channel rises and the river overflows its banks. When the flood waters spill onto the flood plain they slow down as the water becomes shallower and meets more friction from the land. This loss of energy results in the river depositing a fine layer of alluvium onto the floodplain. Over time, the floodplain gets higher as more material is deposited. Floodplains are often used for farming because the alluvium, which is made up of the sediment deposited during the flood, is very fertile.

Estuaries are the tidal mouths of rivers and are found where a river flows into the sea. The characteristics of estuaries are that they are wide, deep and muddy with a mixture of fresh river water and salty sea water. When a river reaches the coast, it is at its widest and deepest stage and has the erosive power to form a large estuary.

An example of a river valley in the UK which has major landforms of erosion and deposition is the river Rheidol in west Wales. The Rheidol is 30 km long and flows west from its source 350 metres high in the Cambrian Mountains to its estuary at Aberystwyth. The valley has a high annual rainfall of over 1,000 mm a year and a medium sized drainage basin of 189 km². The river flows 8 km south from Nant-y-moch to Ponterwyd; then south-west to Aberystwyth.

In the upper valley of the Rheidol there are several fluvial erosional landforms. For example, at Devil's Bridge there is a deeply incised gorge and the Mynach Falls, a spectacular waterfall. Further along the valley near Aber-Peithnant there are some interlocking spurs where the river is narrower, shallower and slower.

In its middle course, there are some fluvial landforms resulting from both erosion and deposition. As the Rheidol flows westwards for 10 km, the valley widens to 600 metres and there are large meanders which occupy the whole floodplain. There are also several oxbow lakes.

By the time the Rheidol reaches its lower course, its river channel has become wider, deeper and faster because it has been joined by tributaries such as the Llywernog, which means it contains a higher volume of water. In this mature stage, there are landforms which have resulted from deposition such as levées. The Rheidol has a wide flat floodplain and in Aberystwyth there are levées that have been artificially raised in order to prevent flooding. At its small tidal estuary, the river finishes its course and flows into the sea.

During lengthy periods of rainfall, water levels in rivers rise and flooding can occur. As flooding can cause considerable loss of life and damage to property, management strategies including hard and soft engineering can be used by government agencies, organisations and home owners to reduce the flood risk, protect homes, and river landscapes from the effects of flooding.

There are several physical and human factors which increase the risk of flooding. The physical factors are the amount of precipitation, the geology, the vegetation and the relief of the land. The human factors include the land use of the drainage basin for example the amount of settlement, and the extent of economic activities such as farming and forestry.

How do physical factors such as precipitation, geology, and relief affect the flood risk? Precipitation is the moisture falling from the atmosphere – as rain, hail, sleet or snow. If the amount of precipitation is low and spread over a long period, there is a longer lag time because infiltration, throughflow and groundflow can take place. This means that it will take longer for the water to reach the river, reducing the likelihood of a flood occurring. However, if there is a large amount of precipitation over a prolonged period of time the soil can become saturated and water can no longer soak into the ground. There is a shorter lag time as there is increased surface runoff into the streams and rivers resulting in the river levels rising quickly and flooding occurring. In the spring in some places, there is an increased risk of flooding when the snow melts, and meltwaters swell the volume of water in the rivers.

Geology is another important physical factor influencing the amount of flooding. If the rocks and soils of the drainage basin are impermeable there is a higher risk. This is because there is less infiltration and more surface runoff. Flooding is more frequent in the granite areas of Cornwall and around Tunbridge Wells in Kent where there are clay soils.

If the relief of a drainage basin includes steep slopes this will result in greater overland flow and a shorter lag time. After a period of heavy rainfall, the surface runoff is rapid and excess water reaches the rivers quickly resulting in higher and more frequent floods. This happens at Cockermouth in the Lake District where not only is the town surrounded by steep slopes but also there is the confluence of two rivers.

Where the gradient is less steep, surface run off is slower, more infiltration occurs and there is a longer lag time. However, some low-lying areas such as the Somerset Levels and Lincolnshire are particularly at risk from flooding because the valley sides are lower and the river banks are more easily breached.

How do human factors such as land use affect the flood risk? There are human activities that increase discharge and lead to more frequent floods. These include building on flood plains, poor farming practices and clearing woodland. Concrete and tarmac surfaces are impermeable therefore rivers in urban areas have shorter lag times, higher peak discharges and flashier hydrographs because the rainwater reaches the drainage systems quicker. Poor agricultural practices – such as ploughing up and down slopes – can result in less infiltration and more rapid surface runoff into the streams and rivers, which are then more likely to flood.

How can the use of hydrographs help to show the relationship between precipitation and discharge? A hydrograph is a special graph which shows the relationship between the discharge of a river and the amount of rainfall that has fallen within the drainage basin over a specific period of time. On a hydrograph, the rising limb shows the increase in river discharge as rainwater flows into the river and the falling limb indicates the decrease in discharge as the river returns to its normal level.

Discharge is the volume of water in a river that flows past a given point on a stream or river bank within a given period of time. River discharge is measured in m^3 per second or cumecs. After a period of rain, the volume of water in a river rises until it reaches its highest level which is its peak discharge. There is a delay between peak rainfall and peak discharge because most rain doesn't fall directly into the river channel but falls on the surrounding land and makes its way to the river. This delay in reaching the river channel is called the lag time.

The shape of a hydrograph is determined by the time it takes for the rain water to reach the river. If the rain cannot soak into the ground because the land is concreted over or is already saturated the water runs off over the surface and quickly reaches the river. This results in a flashy storm hydrograph which has steep rising and falling limbs. If the ground isn't saturated and is permeable, infiltration and throughflow occur and it takes longer for the water to reach the river. This results in a subdued hydrograph which has gentler rising and falling limbs.

If deforestation has occurred there is an increased flood risk as there is less interception and more surface runoff resulting in a shorter lag time and a flashy hydrograph. In contrast if the drainage basin is heavily wooded the lag time is increased and the flood risk reduced as there is more infiltration and interception as the trees and their roots trap and absorb the rainfall.

So, even if drainage basins have received the same amount and pattern of rainfall over a period of four hours, their hydrographs can look very different. The drainage basin with steep slopes, little vegetation and an impermeable rock will have a flashy hydrograph with a steep rising limb; whereas the drainage basin with gentle slopes, woodland and a permeable soil will have a subdued hydrograph with a gentle rising limb.

A flood occurs when river discharge exceeds river channel capacity and water spills out of the channel onto the floodplain and other areas. Flooding can be controlled by hard engineering such as building dams and reservoirs, straightening, embankments, and flood relief channels. Each strategy has its own costs and benefits. Flood risk is the predicted frequency of floods in an area.

Different management strategies can be used to protect homes and property from the effects of flooding. Hard engineering involves the building of entirely artificial structures using various materials such as rock, concrete and steel. These structures reduce, disrupt or stop the impact of river processes. Soft engineering uses the natural environment surrounding a river, using schemes that work with the river's natural processes. Soft engineering is usually much cheaper and offers a more sustainable option as it does not interfere directly with the river's flow.

Hard engineering solutions include building dams and reservoirs, straightening rivers, building embankments and flood relief channels. Each strategy has costs and benefits.

Dams are man-made barriers made of earth, concrete or stone which are built across a river to interrupt river flow and to hold back the flood water. Reservoirs are man-made lakes which form behind dams. They store water and control the discharge of the river. Kielder Water in Northumberland which was opened in 1982 is the largest dam and reservoir in the UK.

Building dams and reservoirs is expensive. For example, Kielder Water cost £167 million to build in the 1970s. It would be more today. The costs include the loss of farmland and homes when valleys are flooded to create reservoirs. Over 1.5 million trees and 2,700 acres of farmland were flooded when Kielder Water was built. Other problems caused by building dams and reservoirs include disruption to wildlife habitats, the displacement of people, and the build-up of algae in the reservoir, which can deoxygenate the water. In addition, there are high maintenance costs as the build-up of sediment behind the dam could lead to structural failure with potentially catastrophic effects downstream.

The benefits of dams and reservoirs include a reduced risk of flooding, the provision of irrigation and drinking water all year round, and the use of the reservoir for fishing, recreation and hydro-electric power. In addition, the leisure and tourism industries benefit from the increased opportunities for fishing, sailing and recreation. Other benefits are the prestige from having a large multi-purpose scheme, the protection of farmland downstream, and improved navigation. Although these hard engineering works are expensive to build and maintain, they are effective, reassure people and reduce insurance premiums.

River channel straightening is another hard engineering strategy which reduces flood risk. Straightening rivers involves removing most of a river's meanders by building artificial cut-throughs. This enables the water to flow faster downstream because there is less friction with the bed and banks. There is less chance of flooding because the rising level of water in the river is contained within the river channel and the river is less likely to break its banks.

The capital costs of river straightening are medium to low – roughly in the region of £200 a metre. However, building straighter channels shifts the problem of flooding downstream and causes the water to flow faster which leads to more erosion and less deposition. This can have negative impacts on the river ecosystem. If the river channel has been lined with concrete this can be visually unattractive and deprive burrowing river bank animals of their habitat.

The benefits of river straightening are that it is a small-scale approach, inexpensive and provides protection to built-up areas. As the straightened river is shorter, navigation is also improved.

Embankments, which are artificial levées along the side of the river designed to make the river deeper, are another hard engineering solution. Embankments enable the river to hold more water as they act as a dam. They are usually covered with grass but can be built of concrete if the flood risk is high. Often, they are steeper on the river channel side than the land side and can be raised further if need be.

Embankment costs are medium and depend on their size, shape and length. Typically, they cost £4,000 a metre to build and have medium maintenance costs. However, they can be breached and therefore are not 100% reliable. Where harder materials such as concrete and gabions are used to build them, habitats can be damaged and the appearance of the river valley negatively impacted.

The benefits of embankments are that they protect built-up areas and allow the floodplain to be built on. Embankments made from earth can also be positive in terms of providing habitats for riverbank animals such as water voles, kingfishers and otters.

Flood relief channels are artificial passages built close to rivers which flood often. These overflow channels reduce the risk of flooding because excess water can flow into them when river levels rise too high. Homes and property can be protected from flooding as flood water can be diverted to an unused area or into another river. An example of a flood relief channel is the Jubilee River, which acts as an overflow channel for the River Thames in London. When water levels in the Thames are too high, water is diverted into the Jubilee, reducing the risk of flooding.

The costs of building and maintaining flood relief channels are low. However, although flood relief channels lower the risk of flooding, they do not prevent flooding all together. Sometimes the use of these channels can lead to flooding in areas unused to being flooded, which can cause even bigger problems.

The benefits of flood relief channels are that they are effective and inexpensive. Their installation can also create new river ecosystems.

Overall, the advantages of building hard engineering structures are that they remove excess flood water quickly, protect homes and property, and reassure people that the government is doing something about flooding.

Soft engineering schemes such as flood warnings and preparation, flood plain zoning, planting trees, and river restoration, work with the natural fluvial processes, in order to reduce the effects of flooding. They aim to slow down the movement of water into the river channel, and do not involve building artificial structures. Soft engineering is usually much cheaper and more sustainable than hard engineering, as it does not interfere directly with a river's flow, and is more ecologically sensitive. Flooding is treated as a natural event which people are expected to live with.

Flood warnings and preparation help people to take action before a flood occurs. Flood warnings provide reliable advance information about possible flooding and give people time to remove possessions and evacuate areas. In many flood risk areas, people can sign up for flood warnings from the Environment Agency. These alerts warn of the risk of flooding from rivers, the sea and groundwater. There are three levels of flood warning – flood alert, flood warning and severe flood warning. Floodline is a free 24-hour help line that also gives information on the latest flood risk. Residents are alerted by phone, email or text when flooding is expected. They have time to prepare for the flood.

People living in high flood risk areas can prepare a plan about what to do if a flood occurs. This plan can include keeping emergency kits at home and knowledge of where to go to keep safe. If people are properly prepared, the damage caused by a flood can be minimized.

The advantage of using flood warnings and preparation as a strategy are the low costs involved – the government pays for the Environment Agency website and the emergency help line. However, this strategy does not prevent people's homes and businesses from being flooded. Also, people may not always act appropriately, especially if the warnings turn out to be false alarms.

Flood plain zoning aims to minimize the amount of damage that flooding can do to homes and businesses. Land nearest to the river, which is most at risk of being flooded, is reserved for low-value land uses such as pastoral farming and playing fields. High value land uses such as homes, offices and shops are built further away from the river so that they are less likely to be affected by flooding. Floodplain zoning also includes creating places along the river valley where the flood waters can go. These places include new wetlands and water meadows.

The flood plain zoning strategy is a low-cost solution that does not involve building structures. When flooding does occur, there is less damage and fewer insurance claims.

Planting trees is another soft engineering strategy. More trees are planted in a drainage basin in order to increase interception and storage, and reduce surface run off and river discharge. Rivers are less likely to flood if their valleys are heavily wooded. The Environment Agency estimates that planting trees can reduce flooding by as much as 20 % in a drainage basin.

Planting trees is a low-cost solution although actual planting of saplings is labour intensive and does result in the loss of farmland. The appearance of the countryside changes from grass to woodland which some people may consider to be less aesthetically pleasing. Woodland needs maintenance but can also be used for leisure and recreation.

The benefits of planting trees are that there are many positive environmental impacts. Woodland creates new habitats for animals and increases biodiversity in an area. Trees improve water quality by filtering pollutants out of rainwater. Trees improve air quality and help to reduce greenhouse gases in the atmosphere by absorbing carbon dioxide. In addition, woodland reduces soil erosion.

River restoration is another type of soft engineering. It involves restoring a river to its natural channel and course. To do this, concrete structures such as weirs and embankments are removed. In their place, new meanders and islands are created to lengthen the course of the river and to slow down the velocity of the water. Trees are also planted and wetland areas created to slow down the surface run off and to reduce the risk of flooding.

There are disadvantages to river restoration as this strategy does not prevent flooding and damage to people's homes and businesses may still occur. There is less land available for housing which can be a problem in densely populated areas. However, river restoration is good for the environment and can increase the size of parks and recreational areas as well as enabling fishing to take place.

An example of a successful river restoration scheme is the River Quaggy in Greenwich. This river had previously flowed through underground drains but in 2007 was brought back to the surface and restored close to its original meandering course. The river valley was planted with wildflower meadows and avenues of trees which created new habitats and parkland for the local people.

Overall, the benefits of soft engineering are that the river is left in its natural state and is more sustainable as the fluvial processes and ecosystems are enhanced and improved. It is cheaper than hard engineering options. However, the flood risk has only been reduced and not removed completely. In addition, if a river changes course this may affect existing roads and buildings which can be expensive to remedy. So, soft engineering may not be suitable for places which are already built up and where there is a shortage of land for homes and services.

An example of a flood management scheme in the UK is found at Boscastle in Cornwall. The scheme was required after the village was badly flooded in August 2004, following a torrential downpour. Over 75 mm of rain fell in just two hours. The water flowed into the rivers Valency and Jordan which join just above Boscastle. River levels rose rapidly and two billion litres of water rushed down the Valency valley through the middle of Boscastle.

Although there was no loss of life due to the well-coordinated rescue operation, the floodwater caused a great deal of damage to property and to local wildlife habitats. There was coastal pollution as debris and fuel from cars flowed out to sea. The clean-up and rebuild operation caused long-term disruption to the village as well as a loss of income from tourism.

After the flood, both hard and soft engineering management strategies were adopted to reduce the risk of future floods. The £4.6 million flood defence scheme was completed in 2008. The hard engineering that took place included widening and deepening the river channel of the Valency, so that the river could hold more flood water, without breaking its banks. The car park was relocated further from the river and was raised and covered with a permeable surface. These measures were undertaken in order to prevent parked cars being washed down the river into the sea during another flood. A higher flood wall was built to keep the water in the river. A culvert, which is an underground tunnel, was built to divert water from the river Jordan if need be. The bridges across the Valency were made higher and wider so that more water could pass underneath.

Soft engineering strategies were also used to reduce the risk of future flooding. Some element of flood plain zoning was adopted as the land next to the river was left in its natural state and homes were built further away. More meanders were created to slow down the flood waters. The site of the former car park next to the river was lowered and reinstated as floodplain. Trees that had formerly blocked the flow of the river were removed but new trees were planted to increase infiltration and interception.

The main social issue that the new flood management scheme raised was to do with housing. The homes damaged in the flood were rebuilt but the building of new homes is restricted because of the flood risk. This has resulted in a housing shortage which has increased house prices. While the flood defences were being installed in Boscastle many people were unable to return to their homes for many months which caused some distress and anxiety. However, in spite of several periods of heavy rain the flood defences have protected the village.

The economic issue raised was the cost of building the flood management scheme. The funding came from a variety of sources, including the Duchy of Cornwall, the county council and the Environment Agency. The scheme had to be completed as quickly as possible, so that local businesses could start trading again to make up for the loss of income they had incurred as a result of the flood. The flood defences have to be maintained and money has to be found to pay for this.

The environmental issues the engineers had to consider were how to reduce the risk of flooding but at the same time, how to allow the fluvial processes of erosion, transportation and deposition to take place. They had to balance the need to protect homes and businesses with the desire to protect and enhance the environment. Since the flood defences have been completed there has been an increase in habitats and biodiversity.

13. Glacial landscapes in the UK

I ce was a powerful force in shaping the physical landscape of the UK. In the past, there have been several ice ages, which are periods of time when ice sheets cover the land. Within ice ages there are colder glacial periods and warmer interglacials. Currently the UK is in an interglacial which began 10,000 years ago.

During the last glacial period, the climate of Great Britain and Northern Ireland was so cold that thick layers of ice covered the land. The ice formed over many years as layers of snow built up. The weight of the snow on top compressed the snow below into a firn, which became ice when all the air and space between the snowflakes had been squeezed out.

The ice was a major agent of weathering, erosion, transportation and deposition with enough force to create dramatic landscapes. When the climate warmed and the valley glaciers and ice sheets melted, the land that had been covered with ice, had been changed in many ways.

The maximum extent of the ice cover across the UK during the last Ice Age was two-thirds of the country. At the peak of the glacial period about 25,000 years ago the ice covered the whole of Scotland, much of Wales and most of northern England. The only areas where there were no glaciers or ice sheets were the Midlands, southern England and the extreme south of Wales. When the ice melted, glacial landscapes were left in the Scottish Highlands, the Lake District and North Wales.

Glacial processes, which include weathering, erosion, movement, transportation, and deposition, occur when it is very cold with the air temperature usually below or just above freezing point. The main type of weathering in glaciated areas is freeze-thaw weathering. This process, which is also called frost-shattering, occurs when water enters the cracks in exposed rocks. At night, it is so cold that this water turns into ice. As more ice builds up in the cracks, the mass of ice exerts pressure on the surrounding rock, and makes the cracks larger. During the day, when the temperature rises the ice thaws only to freeze again when the temperature falls. Over time, the constant repetition of this process causes the rocks to be weakened so much that pieces of rock break off.

Glacial erosion includes abrasion and plucking. Erosion is the most powerful glacial process, especially during very cold long winters. Abrasion occurs when rocks and boulders frozen into the base and sides of a glacier are scraped against the rocks along the valley. The frozen rocks act like a giant file, scratching and scraping the rocks below in a process similar to using sandpaper. The rocks along the valley sides and base are polished until much smoother. The ridges and furrows which are left on rock surfaces are called striations.

Plucking is a type of erosion where melt water freezes and bonds the glacier to the rocks on the valley floor and sides. As the glacier moves downhill, the ice plucks or pulls out large pieces of rocks from the valley floor and sides. These rocks are transported down the slope leaving a rough landscape behind.

Movement, including basal sliding and rotational slip, is another important glacial process. During the summer, when it is warmer and the ice starts to melt, glaciers start to move slowly downhill. The meltwater lubricates the bottom of the glacier making movement easier. The glacier moves faster when it is heavier and when it is going down a steep slope with no debris in the way. In winter, the glacier keeps moving slowly down its valley because the sheer weight of the ice, and the force of gravity, causes the ice crystals to change shape and become more like plastic. When a glacier slides over the rocks the process is called basal sliding.

Rotational slip occurs when ice is trapped in a hollow such as a corrie. Meltwater collects underneath the ice and provides lubrication which enables the glacier to move back and forth in a circular motion. Rotational slip gives a corrie glacier enough energy to flow over its lip, down the mountain side and join the main valley glacier. The repetition of this circular movement deepens hollows and corries into larger bowl shapes.

Transportation is another important glacial process. Glaciers can carry the rocks and sediment frozen within them and to their top and bottom for up to 500 miles. During the summer, meltwater rivers flow out from the glacier's snout and transport large amounts of sediment further down the valley. Glaciers can also push everything in their path including large boulders, small pebbles and mud slowly forward in a process called bulldozing.

Deposition, which is the laying down of sediment, is another important glacial process that shapes the landscape. The sediment that is deposited is called till or outwash. Till, which is also known as boulder clay, is composed of an unsorted mixture of silt, rocks and boulders. The sediment can be all shapes and sizes – round, jagged, large or small. All the till has been eroded from the valley floor and sides by the glacier and carried along until it is deposited as ground moraine over a large area.

Outwash material is chiefly sand or gravel, deposited by meltwater streams in front of, and underneath, a glacier. The material is sorted and rounded by water action. Outwash sediment is finer than till as it has been transported and sorted by the meltwater, before being deposited further down the valley on an outwash plain. The larger heavier rocks are deposited first, followed by the smaller finer particles.

Why do glaciers deposit sediment such as till and outwash? Glaciers deposit sediment when they no longer have enough energy to carry it further downstream. This loss of energy occurs when the flow of ice becomes too slow or the ice melts. This usually happens at the snout of the glacier, which is at the end of the valley, where there is lower land and warmer air.

Distinctive glacial landforms result from the different physical processes. Which landforms result from glacial erosion and what are their characteristics and sequence of formation? Landforms which result from glacial erosion include corries, arêtes, pyramidal peaks, truncated spurs, glacial troughs, ribbon lakes, and hanging valleys.

Corries, which are also called cirques, are armchair-shaped hollows in the mountainside formed by glacial erosion, movement and weathering. Corries are found on the side of north facing mountains. The characteristics of corries are that they are steep-sided rocky hollows with a raised rock lip. The formation of corries takes place in a sequence. During glaciation, snow filled the holes in the ground and turned to ice which remained frozen on the north-facing slopes due to the lower temperatures there. Over time, freeze-thaw weathering and plucking of the sides and back wall made the hollows larger and deeper. The rocks that were eroded, fell down the mountain slopes and formed piles of scree at the bottom. The back and sides of the corrie became steeper and the ice formed a small round glacier. Rotational slip enabled the small corrie glaciers to move out of the corrie and downhill to join the main valley glacier. At the end of the Ice Age, moraine on top of the corries' rock lip trapped meltwater in the corrie forming a small round lake called a tarn.

Arêtes are sharp, knife-like ridges found at the back of corries and between two glaciated valleys. The characteristics of arêtes are that they are very steep sided high narrow ridges. The formation of arêtes occurred when the back walls of corries and the sides of glaciated troughs, were weathered and eroded by the ice backwards towards each other. Through freeze-thaw weathering and plucking the rocks were shaped into ridges, which over time became steeper sided and narrower.

Pyramidal peaks are three- or four-sided mountains, which are found in glaciated uplands, in places where several corries cut back into a mountain and meet at a central point. The characteristics of pyramidal peaks are that they are shaped like a pyramid with a sharp point or horn at the top. The formation of pyramidal peaks occurs when three or more corries were eroded backwards by the ice, leaving only a single high peak in the middle. The rocks were broken up through a combination of plucking, abrasion and freeze-thaw weathering. Relatively few pyramidal peaks are found in the UK because the mountain tops were often covered by ice during the last Ice Age.

Truncated spurs are former river valley spurs. The characteristics of truncated spurs are that they are sloping ridges of land which have a blunt end, steep gradients and convex slopes. These areas of harder rocks jut out alternatively on the opposite sides of a glacial valley. They were formed when valley glaciers eroded the sides of a glacial trough through a combination of abrasion and plucking.

Glacial troughs are former river valleys enlarged during glaciation. The characteristics of glacial troughs are that they are U-shaped with wide, flat floors and high steep sides. The formation of glacial troughs occurred during a glacial period when they were occupied by valley glaciers. The rocks on the valley floor and sides were eroded by glacial processes including abrasion and plucking. Lateral erosion widened the valleys and vertical erosion deepened them. As the valley glacier moved slowly down the valley, further abrasion and plucking occurred. Some glacial troughs are particularly large and deep because they were fed by several tributary glaciers all flowing into the main valley glacier. After glaciation, ribbon lakes and different types of moraines were left in glacial troughs.

Ribbon lakes are deep lakes found in former glaciated valleys. The characteristics of ribbon lakes are that they are shaped like a long narrow ribbon. Ribbon lakes formed in glacial troughs at the end of the last Ice Age when the valley glaciers melted. If the meltwater was trapped behind a barrier of terminal moraine, which prevented the water from flowing away, a lake formed. Different types of moraines are found on the floor and sides of ribbon lakes.

Hanging valleys are the valleys of the tributaries of the original river that was flowing in the valley before glaciation. During the Ice Age, a larger valley glacier eroded the lower valleys of these tributaries away leaving a sharp drop into the main valley. When the ice melted, the smaller tributary valleys were left hanging high above the main valley floor. Waterfalls often mark these hanging valleys as the rivers plunge down into the valley below to join the main river. Alluvial fans of sediment often build up at the bottom of these waterfalls.

Which landforms result from glacial transportation and deposition and what are their characteristics and sequence of formation? Glacial landforms which result from processes such as transportation and deposition include erratics, drumlins, and different types of moraines.

Erratics are rocks and boulders which have been transported and deposited by a glacier some distance from their source region. The characteristics of erratics are that they are large irregular shaped rocks.

Erratics were formed when they were plucked from the frozen ground by a glacier moving over them. They were transported by the ice down the valley and abandoned when the ice melted.

Drumlins are hills or mounds found in clusters on the floor of a glacial trough. The characteristics of drumlins are that they are small, egg-shaped hills with the longer axis parallel to the former direction of ice. Drumlins formed at the end of the Ice Age when the ice melted. Sediments such as glacial till, which were frozen to the bottom of glaciers, was deposited in piles by melting glaciers as they moved down a valley.

Moraines are frost-shattered rocks and material which were eroded from the valley floor and sides, before being transported by glaciers down valleys, and then deposited. There are several different types of moraines, including ground, lateral, medial and terminal. The characteristics of moraines are that they are made up of different types, shapes and sizes of glacial material, which has been eroded from the landscape. Moraines formed when the ice melted and the sediment the ice was carrying was deposited.

Ground moraines are made up of rocks and boulders which were trapped underneath glaciers. When the ice melted this material was dropped in lumps and heaps. Lateral moraines are found at the side of valleys and are often mixed up with scree. They formed at the sides of a melting glacier. Medial moraines are ridges in the middle of a valley. They were deposited when a valley glacier melted. Terminal moraines are found at the end of the valley and mark the last position of the snout of the glacier. They often consist of large boulders that the ice pushed along. When the ice retreated, these boulders were left at the furthest point that the ice reached.

An example of an upland area in the UK affected by glaciation is Cadair Idris, a mountain 852 metres high in the Snowdonia National Park in north Wales. It is a north facing escarpment and an ancient sill made from hard intrusive igneous rocks. This glaciated upland area was significantly changed by glacial erosion and deposition during the last Ice Age.

During glaciation, snow and ice filled the hollows on the north facing side of Cadair Idris and developed into at least five corrie glaciers, the largest of which is Cwm Cau. Cwm Cau corrie faces east and is over 500 metres wide. It has a steep 80° back wall, which is 500 metres high and is made from hard lava and ash. The sides were eroded back into a zone of softer mudstone and therefore the corrie basin is elongated rather than the more typical circular shape. Cwm Cau contains a lake called Llyn Cau, which is 20 metres deep at the front and 40 metres deep at the back. A small stream Nant Cadair flows from the lake.

Cadair Idris has a number of arêtes, which are popular with walkers as they provide a natural route way and from which there are excellent views. At the back of Cwm Cau is an arête called Craig Cau – it is over 800 metres high. It formed during the Ice Age when Cwm Cau and Cwm Gadair cut back towards each other.

Penygadair at 893 metres high, is the highest peak in the Cadair Idris range, and is a pyramidal peak formed due to the backwards erosion of three corries, including Cwm Cau and Cwm Gadair.

The 7 km long valley of the Afon Dysynni, south of Cadair Idris and Dolgellau, is a glacial trough. It has a 500 metres wide flat floor and steep sides. At Cadair Idris, truncated spurs are located in the glacial trough of the river Dysynni at Tal-y-llyn and near Dol-ffanog. The Nant Cadair flows in a hanging valley which falls 200 metres in a series of waterfalls down to the main Tal-y-llyn valley.

At Cadair Idris, a valley glacier occupied the valley of the Afon Dysynni which followed the Bala Fault, a major line of weakness which was more easily eroded. When the ice melted a ribbon lake called Tal-y-llyn Lake formed behind a large post-glacial barrier. This barrier was created when a large amount of rock and moraine slumped down from the arête Graig Goch at Tal-y-llyn, a small hamlet at the western end of the lake. Today the lake is 1.5 km long and 50 metres wide although it was formerly much wider and deeper. Various streams flow into the lake forming alluvial fans and deltas on the valley floor. The upper end of the lake is becoming shallower as a result of the fluvial deposition.

Erratics are found in the Nant Cadair valley. They are made from hard volcanic rocks and have been deposited on top of softer mudstones. A deep layer of till, covering large areas of Cadair Idris, was left when the ice melted. Moraines are also glacial deposits and are similarly made up of material eroded by the ice from the valley floor and sides.

There is a drumlin south west of Tal-y-llyn in the glacial trough of the river Dysynni. It is 2 metres long and 10 metres high. Ground moraines can be found near Tal-y-llyn and low lateral moraines, 3 metres high, are found around the rim of Tal-y-llyn Lake. Medial moraines have disappeared from the Cadair Idris area but there is a terminal moraine at the western end of Tal-y-llyn.

How do glaciated upland areas provide opportunities for different economic activities, and what management strategies can be used to reduce land use conflicts? Glaciated upland areas provide limited opportunities for economic activity as they can be extreme environments with their steep slopes, thin soils, low temperatures and heavy rainfall. However, economic activities that are possible include tourism, farming, forestry, quarrying, hydro-electric power, renewable energy, water supply, and military training. Tourism is one of the most important economic activities in glaciated upland areas. This is because these landscapes are beautiful, rugged and offer many opportunities for outdoor activities such as walking, cycling, sailing and kayaking. Tourism is a major source of income in the glaciated areas of north

Wales, Scotland and the Lake District. People visit these places to enjoy the mountainous landscape created by glaciation.

Farming is another important economic activity in formerly glaciated areas. Farming has been a traditional way of life for centuries even though upland soils are thin and acidic and the weather can be extreme. Pastoral farming, especially sheep grazing, is the main type of farming practised because sheep can tolerate the poor vegetation and the cold, wet and windy conditions. In some highland areas cattle and deer are also reared.

In the U-shaped glacial troughs, the land is flatter and the soils are deeper due to the deposition of moraines. This makes some arable farming possible with typical crops including cereals and potatoes, and grass for winter feed. Lowland glaciated areas are often covered by a thick layer of till and boulder clay, which is very fertile. However, as farming incomes are low, many farmers have diversified into other economic activities such as holiday cottages.

Forestry is one of the few economic activities found in places with acidic soils and steep slopes. Coniferous trees are planted for timber for use in the construction industry and for making paper. Forestry provides employment and contributes significantly to local economies in upland areas.

Quarrying is another economic activity in upland glaciated areas found where the rocks are hard and resistant. In north Wales there are several slate quarries, for example at Penrhyn and Dinorwic. Slate is quarried to be used for roofing, flooring, worktops and headstones. When the ice sheets melted in lowland areas outwash sands and gravels were deposited. Today sand is quarried for use in the construction industry at Leighton Buzzard in Bedfordshire.

What conflicts occur between different land uses in glaciated uplands? Disagreements can occur between the different users of an upland area about how the land should be used. For example, landowners who use their land for farming or forestry might want to restrict the number of visitors to the area. Disagreements can develop into a conflict if a satisfactory compromise cannot be found.

Tourists can come into conflict with farmers if the countryside code is not followed. For example, visitors can damage footpaths, leave gates open, drop litter, frighten farm animals and generally cause a nuisance. Large numbers of cars arriving at the same time on a Bank Holiday Monday can lead to congestion, air pollution and a shortage of parking in rural areas.

Forestry can cause conflict with local residents if the heavy trucks used to transport the logs leads to traffic congestion, noise pollution and the compaction of the ground. The risk of flooding can increase when the trees are felled. Quarrying can scar the landscape and may lead to water pollution. Tourists are less likely to stay near quarries which are noisy industrial sites.

What conflicts occur between development and conservation? Economic development of upland areas can lead to new jobs and opportunities for the local people. However, any new economic activity can lead to conflict with conservationists who manage the fragile upland environment in order to preserve, protect or restore it. The tourist industry can clash with conservation if areas of natural beauty are built on or if nature reserves are damaged by noise, trampling and fires. Forestry can also lead to conflict if conservationists object to commercial plantations with trees in rows, which impact on the local wildlife and ecology. However, it is possible to have both development and conservation in the uplands if the appropriate management strategies are deployed to mitigate the negative impacts. Economic development can also increase the amount of money available to spend on conservation.

What is an example of a glaciated upland area in the UK used for tourism to show the attractions for tourists, the social, economic and environmental impacts of tourism, and the strategies used to manage the impact of tourism?

Snowdonia National Park in north Wales is an example of one such area, as it is a glaciated upland area with an important tourist industry. After the Ice Age, the deep ice sheet, which had covered the mountains, melted exposing the beautiful landscape of today. Tourism is now an important economic activity in Snowdonia. Every year over three million people visit the National Park, attracted by the many natural and built attractions, the beautiful landscape and the range of activities on offer.

There are several spectacular landforms in Snowdonia including glacial valleys and mountains. In addition, visitors can enjoy the many varied landscapes such as moorland, waterfalls, woodland and hill farms.

Mount Snowdon (Yr Wyddfa in Welsh), at 1,085 metres above sea level, is the highest mountain in England and Wales. This pyramidal peak is visited by over 700,000 people a year.

To the north west of Snowdon there is Llanberis Pass, which is a glacial trough and Llyn Padarn, which is a ribbon lake. A corrie popular with visitors is Glaslyn, which is 600 metres high up on the north east slope of Snowdon. Many walkers traverse the Snowdon Horseshoe following Crib Goch arête, which is 921 metres high, and Crib-y-Ddysgl.

Activities popular with tourists include hill walking, abseiling, gorge scrambling, rock climbing, water sports, fishing and mountain biking. There are many nature trails and wildlife to be spotted including buzzards, otters, polecats and fallow deer. Many visitors take a train ride to the summit of Snowdon to enjoy the spectacular scenery and views.

There are many tourist villages in Snowdonia. Betws-y-Coed is a busy village located at the confluence of the rivers Conway and Llugwy where there are gift and tea shops, historic bridges, walks and a National Park visitor centre. Llanberis is a popular tourist destination at the foot of Snowdon.

Visitors enjoy the medieval castles including Harlech Castle which was built by Edward I during his invasion of Wales in the 13th century. There are several disused mines and quarries which are now open to visitors. At the Llechwedd Slate Caverns near Blaenau Ffestiniog tourists can see the vast caverns and an underground lake as well as learn about the history of slate mining. At Sygon Copper Mine there are caverns with stalagmites and stalactites and a visitor centre, and at Penrhyn Quarry there is the longest and fastest zip line in Europe. Heritage railways, which were built for the mining industry have been restored and now run sight-seeing trips. For example, the Ffestiniog narrow gauge railway travels through 14 miles of captivating scenery using the original 1860s locomotives and carriages.

Tourism has had significant social, economic and environmental impacts on Snowdonia. The positive social impacts of tourism include the preservation of the local culture including events such as the annual National Eisteddfod which celebrates the culture and language of Wales. However there have been negative impacts on the local people. Tourism leads to a loss of peace and quiet particularly in the summer months. The population density is low – under 199 people per km^2 for most of the year – but at the peak holiday times the honeypots such as Betws-y-Coed and Llanberis become very crowded and congested with traffic jams and overflowing car parks.

Also, not everyone has benefited from the higher incomes that tourism brings – people living in the more remote rural areas have fewer visitors and higher costs. Many families leave and go to places where there are better opportunities. This depopulation leads to the withdrawal of services such as the village school, the local doctor, and rural bus services. This in turn leads to further decline. In addition, although many of the local people speak Welsh English is the language of tourism.

Tourism has many positive impacts on the local economy. Every year, the income from tourism exceeds £300 million. The multiplier effect leads to the creation of jobs and opportunities in hotels, shops, cafes, outdoor pursuits and tours. In 2015, there were 7,000 people employed directly or indirectly in tourism in the Snowdonia National Park. These jobs have increased local wages which averaged £21,000 in 2017. The workers spend their money locally – increasing demand for goods and services such as post offices and buses. Tax revenues increase and the government has more money to spend on public services including education and health care as well as improving infrastructure like roads and sewers. By working in tourism, workers can learn transferable skills in customer service and computers resulting in a more skilled workforce.

There are some negative economic impacts on the economy. Many jobs in tourism are low paid and seasonal. Local people wanting alternative careers in professions such as engineering have to move elsewhere. Economic leakage can occur if the profits from tourism leave the region and if foreign workers send their wages abroad. Many people buy second homes in Snowdonia which can lead to housing shortages and house price rises. Second home owners can bring their provisions with them instead of shopping locally. Prices may rise in shops as tourists have more money to spend and shops cater for the requirements of tourists, not local people.

The environmental impacts of tourism include an increase in water, air and noise pollution. In 2010 untreated sewage entered Llyn Padarn in Llanberis, after a sewer failure caused by an influx of visitors. Motor boats on the lakes can reduce water quality and disturb wildlife. Emissions from visitors' vehicles can reduce air quality and increase greenhouse gas emissions. Noise from cars, lorries and motor boats can disrupt the breeding seasons of wildlife.

Walkers can cause damage to footpaths in the most popular areas. Trampling on the ascent to Snowdon has led to the more fragile plants disappearing to be replaced by more robust grasses. Visitors can also leave large amounts of litter behind, which can contaminate rivers and lakes with micro plastics. Ecosystems and habitats can be destroyed by building new tourist attractions and roads. However, money from tourists can be used to conserve and improve habitats and to conserve fragile environments. Overall, tourism in Snowdonia has brought economic advantages but also environmental problems and significant conflicts of interest.

Snowdonia National Park uses several strategies to manage these land use conflicts and to reduce the negative impacts of tourism. For example, visitors are expected to follow the countryside code which emphasises respecting other people, protecting the environment and enjoying the outdoors whilst staying safe. Visitors are asked to leave gates and property as they find them, follow paths and to take their litter home. They have to keep their dogs under control and to plan ahead so that they are prepared for bad weather and emergencies. Information for visitors on signposts and websites is in both Welsh and English so that the Welsh language is protected.

An integrated transport system is being introduced to Snowdonia in order to limit traffic congestion at peak times. On busy weekends, a park-and-ride scheme reduces traffic jams along the A5 and the A55 main roads. Visitors can use the Snowdon Sherpa, which is a cheap shuttle bus that travels around the foot of Snowdon, linking the six main footpaths up to the summit with the car parks, villages and tourist attractions in the area. To reduce the number of cars, the PlusBike scheme at railway stations enables people to arrive by train and to continue by bike.

New visitor attractions, such as the adventure park at Penrhyn Quarry, are open all year round, which creates more permanent fulltime jobs in tourism.

Enterprise zones have been set up in Snowdonia so that the economy does not become too dependent on tourism. These focus on energy, IT, and the environment. At the Snowdonia Aerospace Centre in Llanbedr, drones are being tested and there are proposals to build a small nuclear power station at Trawsfynydd.

Upland footpaths on Snowdon are repaired routinely, so that the effects of trampling are reduced. As the local stone is protected, stone from other places is flown in. This means that repairing footpaths in remote, steep areas can cost over £180 a metre. Vegetation is reseeded to reduce the visual impact of the erosion and visitors are encouraged to use alternative routes by providing signposting or fencing.

Money from tourism has been used to halt the decline in biodiversity by restoring natural habitats. For example, the Migneit moorland in Upper Conwy has been improved by building small dams in the ditches which drain the moors. This has led to an improvement in water quality and carbon storage, and has enabled species like the Golden Plover to flourish. In other places, work is ongoing to stop the spread of non-native species such as knotweed, which threaten the native flora and fauna.

New buildings have to comply with local planning rules and regulations so that their impact on the environment is minimised. For example, the Snowdon Summit Visitor Centre, which opened in June 2009 and receives 500,000 visitors a year, was built to blend in with the landscape. This low-rise building can withstand the extreme weather found on top of the mountain and is clad in oak and granite. From its large windows there are views over to the Lake District.

These strategies have been partially successful as some of the negative impacts of tourism have been reduced. However, microplastic contamination is a growing problem, and the increase in visitor numbers has resulted in long queues to reach the summit and more work for the mountain rescuers.

14. Challenges in the human environment

T his unit – the challenges in the human environment – is concerned with human processes, systems and outcomes. A process is a series of actions that are done by people in order to achieve a particular end, for example to grow food. A process can also be a natural series of changes, for example the growth of a country's population.

In this unit the human processes which are studied include urbanisation and migration, urban growth and change, economic development, and the provision of resources.

A human system is a set of things that are connected. In human geography there are various systems including energy, transport, industry and agriculture. All these systems have in-puts, processes and outputs. The outcomes of these systems include the food we eat, the clothes we wear and the homes we live in.

These processes, systems and outcomes illustrate the ways in which human activity has changed the environment. The changes have occurred both spatially and temporally. This means that human processes such as urbanisation have taken place in different places at different times.

In this unit, a variety of places and countries, which are at various stages of economic and social development, are used to illustrate the various concepts and themes. In addition, a range of places with varying income levels and scales – including local, national and international are considered. Globally, there are higher income countries (HICs), lower income countries (LICs) and newly emerging economies (NEEs). In addition, there are middle income countries (MICs). This subdivision of countries is based on the World Bank income classifications (GNI per capita), which in 2022 were Low Income $1085 or below, and High Income $13,205 or above.

Newly emerging economies (NEEs) are countries that have begun to experience higher rates of economic development, usually with higher levels of industrialisation. They differ from LICs in that they no longer rely primarily on agriculture, have made gains in infrastructure and industrial growth, and are experiencing increasing incomes and high levels of investment. Examples of NEES include Brazil, Russia, China and South Africa – the so-called BRICS countries.

One of the aims of the challenges in the human environment unit is to develop an understanding of the factors that produce a diverse variety of human environments. Other aims are to examine the dynamic nature of human environments that change over time and place, and the need for their sustainable management. A final aim is to develop an understanding of the areas of current and future challenge and opportunity for these environments.

15. Urban issues and challenges

A growing percentage of the world's population lives in urban areas. Rapid urbanisation is a feature of many low-income countries (LICs) and newly emerging economies (NEEs). Urbanisation is the process which leads to a higher proportion of a country's population living in towns and cities rather than in the countryside.

In 1950, 30% of the world's population lived in urban areas. By 2022 that number had increased to 57%. This trend is expected to continue, with the world's urban population predicted to reach 70% by 2050. As the number of people living in urban areas increases, the percentage living in rural areas will carry on falling.

Some parts of the world are more urbanised than others. Europe and North America both have high levels of urbanisation as over 75% of the people live in towns and cities. South America and Australasia are also highly urbanised. However, Africa is less urbanised as only 44% of the people live in towns and cities.

The global pattern of urban change is uneven. In the last fifty years, Asia, South America and Africa have experienced faster rates of urbanisation than North America, Europe and Oceania. For example, between 1950 and 2020, South America experienced rapid urbanisation, when the percentage of the population living in towns and cities grew from 41% to 81%. However, the urban population of Oceania, in the same period, only increased from 63% to 68%.

Urban trends in different parts of the world, including in the higher-income countries (HICs) and in the LICs, vary. The trend in HICs is for the process of urbanisation to slow down or even to go into reverse; whereas in LICs urbanisation is still occurring at a rapid rate each year. For example, in Japan (a HIC) where 92% live in urban areas, the urbanisation rate is -0.4% per year. Middle-income countries also have slowing rates of urbanisation. For example, in Mexico where 80% of the people now live in towns and cities, the rate of urbanisation has fallen to 1.4% a year. In contrast, urbanisation is occurring at a faster rate in low-income countries. For example, in 2020, only 25% of the population of Uganda (a LIC) lived in urban areas, but the rate of urbanisation was a high 5.6% a year.

One of the factors affecting the rate of urbanisation is migration, as explained by the push-pull theory. Migration is the movement of people from one place to another. There are different types of migration. For example, rural-to-urban migration occurs when people leave the countryside to move to the cities. Another factor affecting the rate of urbanisation is natural increase, which is the birth rate minus the death rate of a population.

What is push-pull theory? It is a model which explains why people move. It outlines the push factors which motivate people to leave the place they are living in, and the pull factors which are the attractions of the place they move to.

The push factors that encourage people to leave their homes include political, economic, social and technological reasons. The political reasons include war and conflict, which often means that people are forced to flee for their lives. The economic factors can include a shortage of jobs and opportunities, sometimes caused by mechanisation of agriculture, or a natural disaster such as a flood or famine. The social reasons can include inadequate or poor health care, housing and education. The technological factors include poor services such as electricity, transport and communications.

The pull factors that encourage people to move to urban areas include political, economic, social and technological reasons. The political factors which lead people to move to cities include sanctuary, peace and security. The economic factors include jobs and new opportunities, higher incomes, and better standards of living. The social factors include improved housing, education and health care, as well as being able to join friends and family already settled there. The technological factors include more reliable electricity supplies, better transport and communications.

Faster rates of urbanisation are occurring in low-income and in lower-middle income countries because of the higher rates of natural increase in the cities. This is due to the better healthcare found there. This results in lower rates of infant mortality and higher life expectancy. In addition, city dwellers tend to be younger adults of child bearing age. In Bangladesh, a middle-income country (MIC), the population of its capital Dhaka grew by over 3% in 2021, even though its overal population increase was only 1%. Rural-to-urban migration is also happening at high levels in other low-income countries.

High-income countries are experiencing slower rates of urbanisation because they started urbanising a long time ago during the industrial revolution. Their cities have already grown to a large size and most of their populations already live in urban areas. Increasingly, the factors encouraging people to live in cities are disappearing as life in the countryside improves, as better services and transport are provided. There is a growing trend for people to move back to the countryside where they can have a better quality of life. Working from home also makes life in the countryside more economically possible. As a result, there is less rural-to-urban migration but more urban-to-rural migration.

Natural increase in high-income countries has also slowed. Birth rates have fallen due to better access to contraception and ageing populations, resulting in slower urban growth. For example, Japan experienced a -0.5% fall in its population in 2021. Tokyo, the capital of Japan, had a 0.18% decline in population to 37 million in 2022.

The emergence of megacities occurred when some cities grew so big that their populations exceeded ten million. The first cities to do this were New York and Tokyo in the 1950s. Since then, many more cities, especially in Asia and Africa, have grown in size and have become megacities due to a combination of factors. These factors include high rates of rural-to-urban migration, the higher rates of natural increase as well as the growth of industry.

The global pattern of megacities is uneven. Today most of the world's 33 megacities are located on coastal plains in the northern hemisphere. Six are in Central and South America and 22 are in Asia. China has six megacities including Shanghai and Beijing. In Europe, London and

Istanbul are megacities as is Los Angeles in North America. There are no megacities in Oceania.

Several of the fastest growing megacities are in Africa. For example, Kinshasa, a megacity and the capital of the Democratic Republic of the Congo (a LIC), is growing at the rate of over 4% a year, due to rural-to-urban migration and natural increase. The push factors from the countryside include widespread poverty and instability, whereas the pull factors are better jobs and opportunities in the capital. In 1984, 2.6 million people lived in Kinshasa but by 2022 this had risen to 15.7 million. It is predicted that this megacity will have 35 million residents in 2050, and 83 million in 2100.

Urban growth creates both opportunities and challenges for cities in LICs and in newly emerging economies (NEEs).

The opportunities that urban growth creates are political, economic, social and technological. When people move to cities, they meet a wider range of people who have migrated from different parts of the country or region. This creates political opportunities as barriers between different groups of people begin to break down as they get to know their neighbours. This can lead to less conflict and fewer disputes between different groups.

The economic opportunities created by urban growth arise when new people move in bringing with them new ideas, tastes and traditions. Entrepreneurs start up new businesses such as opening new shops and restaurants selling different types of food and drink. The increased workforce can enable businesses to open factories and for new industries to be started.

Urban growth creates social opportunities for the new migrants. The newcomers have better access to education and health care in the cities, so there is a rise in levels of literacy and numeracy, and an increase in life expectancy.

Finally, there are technological opportunities created by urban growth – it is easier and cheaper to provide utilities such as gas, electricity, broadband and water to people when they live near to each other in an urban area, than to people living in isolated homes in the countryside.

Overall, urban growth enhances the multiplier effect. If there is an increase in population there is greater demand for goods and services, which in turn creates jobs and opportunities, higher incomes and tax revenues, enabling the city government to improve infrastructure and public services.

However, if a city grows too quickly there can be political, economic, social and technological challenges. The political challenges include establishing an effective local government at a time when there are increasing numbers of inhabitants. If the city council struggles to collect enough tax to pay for services this can lead to dirty streets, poor waste collection and a shortage of school places. If there are not enough police to keep law and order this can lead to bribery and corruption, as well as anti-social behaviour. If there are few planning regulations and poor enforcement of those that exist, unsafe homes and factories can be erected in unsuitable places. This can lead to poor housing conditions, fires, high levels of building collapse, and air and water pollution.

There are also many economic challenges when urbanisation and industrial development occurs rapidly in newly emerging economies. If there are not enough jobs for the new migrants to the city, they will seek casual work in the informal economy, where they are paid in cash and have no employment rights, like redundancy pay and sick leave. Their wages are low because they pay no tax, but the low levels of tax revenue means that the city council cannot afford to build schools and infrastructure, such as new roads, sewers, parks and public amenities.

The social challenges which result from rapid urban growth include shortages of affordable housing. As a result, many newcomers to the city often stay rent-free with friends and family in the unplanned and often illegal squatter settlements which have grown up on waste land. In these informal settlements the housing is poor-quality and make-shift with no mains water or sewerage or proper electricity supply. As the residents pay little or no council tax, there are no basic services provided by the council, such as waste collection and street cleaning. In many slums, there is no police presence so there are problems with gangs, drug dealers and weapons. Thus, many squatter settlements are overcrowded, dirty, smelly, dangerous and unhealthy to live in. However, for many people this is their home and they are reluctant to move away until they can afford to do so.

In addition, health and social care services in a rapidly growing city may be overwhelmed by the increasing number of people needing to be treated and looked after. There may also be a shortage of nurses and doctors, leading to outbreaks of disease and poor health outcomes. Similarly, there may not be enough teachers or schools. Young people therefore enter the workplace with low levels of literacy and numeracy and have to take unskilled work with lower wages and longer working hours. This results in high levels of underemployment and poverty.

Cities that experience rapid growth also have technological challenges. The transport infrastructure may not be built fast enough to keep up with demand. This can result in crowded roads which are poorly maintained and unable to deal with the sheer weight of the traffic. Public transport may also be slow, infrequent and crowded. Many people instead drive their own diesel cars which add to already high levels of air pollution. Pollution is the presence of chemicals, noise, dirt or other substances which have harmful or poisonous effects on an environment.

As a result, many city commuters have to endure poor air quality, traffic congestion, frequent delays and long journeys to work. In addition, it may be difficult in fast-growing cities to provide internet and mobile phone services, as well as a reliable electricity supply.

146

These challenges can be solved by city councils and there is evidence that many squatter settlements and urban areas have improved themselves especially if law and order is enforced.

A case study of Lagos in west Africa can be used to illustrate the location and importance of a major city in a NEE; as well as examining the causes of growth, and explaining how urban growth has created opportunities and challenges. Lagos is a megacity with over 15 million people. It is one of the fastest growing cities in the world. Every year the population of Lagos is increasing at the rate of 3.5% – an extra 500,000 people.

Lagos is a major financial hub with the fourth highest gross domestic product (GDP) in the whole of Africa. There are extremes of wealth and poverty in Lagos. Many billionaires live in luxury in the gated communities on Victoria Island. In contrast, millions of families surviving on low-incomes, dwell in Makoko and the other slum areas.

Lagos is located in the south-west of Nigeria on the Gulf of Guinea coast. The city is sited on the west side of Lagos Lagoon, on a large area of sheltered flat land. Long sandy spits such as Bar Beach protect Lagos from the force of the Atlantic Ocean. To the west is the neighbouring country of Benin and to the east is the Niger Delta where there is a large oil field. Lagos is well connected by roads and railways to the rest of Nigeria. An international airport and a large container port on Tin Can Island provide transport links to other countries.

Starting life as a small fishing village on Lagos Island, Lagos gradually developed into a port. As the city grew, the built-up area spread to the neighbouring Victoria Island and along the main roads and railway lines of the mainland. Three bridges, the Eko, Carter and Third Mainland, connect the two islands with the mainland.

As the city sprawled to the north and west, new desirable suburbs were built in the best locations. Slums developed in the least desirable places such as the swamps. In recent years, land has been reclaimed from the sea to provide more space. For example, Eko Atlantic is a new coastal city being built by private developers on reclaimed land to the south of Victoria Island.

Lagos is an important city regionally, nationally and internationally. It is the largest city in Nigeria and west Africa and has a growing economy. It has several major industries including oil; film and media; and finance. Lagos's coastal location and good transport links attract foreign investment and transnational corporations, for example ExxonMobil. Although Lagos is no longer the capital of Nigeria, it is still the country's largest port and handles large amounts of trade. The city is a leading arts and cultural centre, and is important for education with 11 universities.

Lagos's population is growing because of natural increase and migration. There are many young adults living in the city, so the birth rates are higher than the death rates, leading to population increase. Due to a shortage of jobs in the countryside, many people move to Lagos in search of work. Climate change and unreliable rainfall in the north of Nigeria has led to shortages of food, prompting people to move south to Lagos.

The urban growth of Lagos has created social opportunities for the people who live there. The city has many different types of schools including small and informal ones. Over 68% of children attend secondary school – a higher percentage than in the countryside. However, many people leave school with low levels of literacy and numeracy which makes accessing further education and vocational training more difficult.

There is better health care available in the city than in rural areas as there is a wider choice of clinics and hospitals. In addition, in many parts of Lagos there is clean water and electricity resulting in higher standards of living and a better quality of life.

Lagos also offers many economic opportunities to its residents. The oil industry has led to the multiplier effect as people move to Lagos attracted by the jobs and opportunities it has generated. Some people become entrepreneurs opening small businesses such as hairdressing as they have access to investment and skilled workers in the city. Others find low skilled work in factories or in the informal sector working in street vending, waste recycling – the process of extracting and reusing useful substances found in waste and food delivery, or work in high skilled business and finance jobs in the central business district. Where ever they find work, the residents of Lagos generally have higher incomes than if they had stayed in the countryside. They spend their money on goods and services thus creating a growing home market.

However, rapid urban growth has created challenges in Lagos. When large numbers of people arrive in a developing city the shortage of affordable housing means they often have no choice but to live in a slum or squatter settlement. It is estimated that nearly 10% of Lagos's population live in these informal settlements where there is little or no access to services.

Makoko is a large slum area of Lagos. This squatter settlement is nicknamed the Venice of Africa, because the wooden makeshift houses are built on stilts in Lagos Lagoon. No one knows for sure how many people live there but it is thought to house over 100,000 people. Conditions in Makoko are basic as there is no running water, sanitation, or waste collection. These unhygienic conditions mean that there is a high level of disease.

However, many inhabitants of Makoko have adapted to life in the slum. They travel through the canals by canoe and make a living through fishing and trading timber. Some families have built better homes with satellite TV on the drier higher land. There is a strong sense of community with a range of social activities available. Even though their living conditions are poor, the residents have resisted several attempts to demolish their homes.

Managing the growth of the slums and squatter settlements can be challenging. The rate of population growth is so high that the problems grow faster than they can be addressed. The shortage of accommodation means that overcrowding is common. One solution would be to build new homes elsewhere in Lagos for the inhabitants to move into. However, this is difficult because of the shortage of land and the cost of development. Developers have built on reclaimed land but often they build expensive luxury homes, similar to the ones being built at Eko Atlantic, rather than the new affordable homes that are needed.

Another problem resulting from the rapid growth of Lagos is the provision of clean water, sanitation systems and energy. The slums and squatter settlements were unplanned and built without sewers or water pipes. As a result, the residents provided their own toilets, which are often unhygienic, smelly and responsible for outbreaks of disease. At night there is often no lighting so the toilets can be dangerous to use alone. People use generators to generate their own electricity. Cables are slung between the houses to bring light and power into people's homes. Clean water is bought in plastic bottles, which are thrown away adding to the build-up of rubbish.

Providing health care to the large number of residents in Lagos is a challenge for the local authorities. There are not enough doctors so there are long waiting lists. Medicines and treatment are not free.

Educating so many children can also be a challenge. One solution has been to build more schools in the squatter settlements. In Lagos Lagoon, the Makoko school provides education for the local children, who arrive by canoe. The pyramid-shaped school is built on a floating platform and is designed to cope with the fluctuating water levels and stormy weather. However, the wooden building collapsed not long after opening and had to be rebuilt. Makoko school is powered by solar panels on the roof and is open all year round, although there are not enough places to accommodate all the children who want to go there.

Another challenge is reducing the high levels of unemployment. One solution has been to provide business loans so that entrepreneurs can start up small enterprises. Another is providing more educational opportunities so that people are better qualified and can apply for higher paid jobs in the formal sector.

Reducing crime is another challenge. Crime rates, including those for burglary, vandalism, drug dealing, assault, armed robbery, corruption, fraud, bribery and cybercrime are all high in Lagos. The high levels of poverty, unemployment, and social exclusion mean that some people resort to crime in order to survive. In many of the slums and squatter settlements, gangs are more powerful than the police. Members of these gangs control many aspects of the settlers' lives. The police are mistrusted as there are some corrupt officers. This mistrust makes it harder for the police service to enforce law and order.

One solution to the high levels of crime has been to introduce community policing so that people work with their local crime prevention officers. The built environment has also been improved so that it is harder for criminals to operate. Measures which have been adopted include street lights, having clear sightlines, landscaping, installing fencing and gates, removing graffiti and abandoned vehicles, and street cleaning.

Lagos city council is also managing the environmental issues caused by rapid urban growth. Waste disposal is a key issue. Piles of rubbish often build up in the streets even though over 10,000 tonnes of waste are delivered each day to the Olusosun landfill site. A community of 5,000 rag pickers live next to the site and sift through it, earning a good living by selling scrap metals, wires and other waste. Container ships also import waste, including electronic waste, from all over the world.

One way of reducing the amount of waste going to landfill is to increase recycling rates. This will save resources as new materials do not have to be found to replace those that are thrown away and will reduce the amount of energy needed to process the waste.

There is a lot of air and water pollution in Lagos. Levels of air pollution are high because vehicles on the roads are older and more polluting. Many factories are unregulated and have high levels of emissions. Industrial areas such as Ikeja have weak air and water pollution controls. Many people still rely on fuelwood or kerosene which leads to high levels of smoke in the air.

Traffic congestion is another problem in Lagos, as there are too many cars for the existing road network to cope with. In addition, the roads are poorly maintained which slows down journey times and has a negative impact on the economy. Every year the problem gets worse, as the citizens of Lagos get richer and can afford to buy their first car – over 40% of the new cars in Nigeria are registered in Lagos. More motorways need to be built, but this requires considerable investment.

Lagos has improved public transport by introducing a new bus service called the Bus Rapid Transit. The fleet of modern buses runs along a designated bus corridor and provides a fast, comfortable and reliable service, which enables commuters to get to work on time and thus improve their incomes and quality of life. A new light railway has also been built, which enables people to get to work faster and to access a wider range of jobs which improves their incomes and lifestyle.

However, in spite of all these challenges, Lagos's GDP is growing at the rate of 3% a year, and the standard of living of most citizens is improving.

An example of how urban planning can improve the quality of life for the urban poor can be found in Rio de Janeiro in Brazil, a newly emerging economy. The Favela-Barrio project was started in the 1990s in order to improve the lives of residents living in the squatter settlements which are also called favelas.

Rio de Janeiro's favelas were built by migrants who could not afford to rent or buy a house in the rapidly growing city. Instead, they set up informal settlements on the sides of steep valleys and on waste land along the railway lines. In Rocinha, one of the largest favelas in Rio de Janeiro, more than 100,000 residents still live in an area less than one square mile.

The squatters built their homes, often many stories high, from waste materials such as wooden planks, cardboard and corrugated iron sheets. The favelas had no services like mains water, sewers, waste collection, street lighting or policing. As a result, the streets were dirty, smelly, and dangerous.

Determined to improve the lives of the people living in the favelas, the Favela-Barrio project replaced the shacks with brick and concrete homes; installed basic amenities such as electricity, running water and sewerage; and gave ownership of their properties to the residents. In Rocinha, over 90% of the homes now have electricity, are brick built, and have running water.

Schools, health clinics, banks, drug stores and a radio station were opened in Rocinha. However, even though residents are better educated, healthier and safer than before, standards of living in Rocinha are still lower than in the wealthier suburbs. Many homes remain crowded, unsanitary, and unsafe. Many children do not attend secondary and tertiary education. Levels of health care remain poor. As a result, there are higher levels of contagious diseases such as tuberculosis and health-related disabilities, lower levels of literacy and numeracy, and higher levels of crime and violence than in the wealthier suburbs of Rio de Janeiro such as São Conrado. It will take more time and investment before life improves for all the residents of the world's fastest growing cities.

Urban change is also occurring in the UK as some towns and cities grow and others decline. This change leads to a variety of social, economic and environmental opportunities and challenges. Urban growth has led to many opportunities. When migrants arrive in a city they bring with them new languages, customs, traditions, food, music and religions. This creates many social opportunities, which are chances for people to improve their quality of life, for instance access to education and health care, for the city. New shops and restaurants open up selling different food and drink. The migrants build their own places of worship which adds to the variety and styles of architecture within the city. The connections that the new migrants have with other parts of the country and the world can also enhance trade and build cultural links with many mutual benefits.

In London, the annual Notting Hill Carnival showcases the musical and culinary influences that the immigrants from the Caribbean have had on the city. In Chinatown migrants have opened restaurants and supermarkets selling Chinese food. The Balti Triangle is an area in Birmingham famous for its curry restaurants run by settlers from Pakistan.

There are many economic opportunities, which are chances for people to improve their standard of living through employment, created by the arrival of migrants from different places. The incoming people are often young adults of working age who fill the job vacancies, pay taxes and enlarge the home market. This leads to the multiplier effect as new jobs and opportunities are created by the growing economically active population.

Integrated transport systems are when different transport methods connect together, making journeys smoother and therefore public transport more appealing. City councils use the revenue raised from public transport and parking charges to invest in improvements in integrated transport systems. Better integration should result in more demand for public transport, and should see people switching from private car use to public modes of transport, which should be more sustainable. Integrated transport systems may also lead to a fall in congestion due to less road users, cut traveling times and provide a cleaner more reliable service. For example, in Bristol the city has grown because of the skilled migrants moving there to work in the new industries such as micro electric and silicon design, and animation.

There are also environmental opportunities created by the growth in the number of people living and working in a city. The increase in council tax revenue can be used to improve air quality, create new parks and gardens, and pay for more street cleaning and landscaping. This urban greening benefits both people and wildlife alike.

However, urban growth in the UK has also led to a variety of social, economic and environmental challenges. The social challenges include housing. In the UK, when urbanisation began during the industrial revolution in the eighteenth century, many people moved to the towns and cities in search of work. To accommodate the growing populations, poor quality homes were quickly erected. Standards of living in the slums remained low until most of them were cleared in the twentieth century, and their inhabitants moved to new homes in the suburbs or new towns.

Today it is still a challenge to provide enough homes for the growing population. One solution is to build new housing on brownfield sites, which is land that has been used, abandoned and now awaits some new use. Brownfield sites are commonly found across urban areas, particularly in the inner city. Developing brownfield sites can be expensive if the land has been polluted and has to be decontaminated. However, in many cities it makes sense to reuse derelict land. This is happening in Alperton in the London Borough of Brent where 6,800 new homes are being built along the Grand Union Canal on a former industrial site.

Another solution is to build high-rise tower blocks in the suburbs. However, this can result in local opposition as people object to low-rise homes being overshadowed by tall buildings. In the London Borough of Ealing, local people have objected to high buildings being built next to roads of Victorian semi and terraced houses, protesting that this will result in a radical change in the character of the suburb.

New housing estates can be built on greenfield sites. These are plots of land that have never previously been built on. Most greenfield sites are found on the edge of a town or city. For example, a new development with up to 5,500 new homes, shops, schools, medical centres and leisure facilities is being planned on a former green belt site on the north-eastern edge of Birmingham at Langley. New settlements such as these should also be well connected to the nearest urban centre by public transport or road in order to prevent the residents becoming socially isolated.

An additional social challenge can arise if immigrants from a particular place live together in clusters within the city and do not mix with other groups or learn the local language. This can lead to community tension if steps are not taken to integrate the newcomers.

The economic challenges of urban change include the effects of de-industrialisation on the inner city. When businesses close or relocate, workers who were employed in the factories or workshops lose their jobs and may struggle to gain another. The loss of income impacts not just on individual families but also on the area of the city which has lost its main economic activity. Shops, restaurants and other services close as their customers can no longer afford to use them. The younger more skilled workers move away in search of better opportunities elsewhere leaving behind the elderly and the disadvantaged.

A circle of deprivation, which is the degree to which an individual or an area is deprived of services, decent housing, adequate income and local employment, and decline sets in as the remaining residents struggle with lower incomes and poorer services. In addition, abandoned buildings and factories may become wasteland, derelict, vandalised and covered with graffiti if left vacant for too long, creating urban blight.

When these inner-city areas are redeveloped, the local community is broken up as the remaining inhabitants are rehoused in other areas.

Another challenge arises when economic growth is not evenly spread across a city, resulting in richer and poorer areas and health and social inequalities. In some of the poorest wards of the UK's largest cities life expectancy and educational attainment is well below the national average. For example, in Aston, a poor area of Birmingham, life expectancy is only 76 years compared to 85 years in Four Oaks, one of the richer suburbs. GCSE scores are lower in Filwood, a poorer part of Bristol, than in Stoke Bishop, one of the more affluent areas.

There are also environmental challenges which arise as a result of urban change. When urban sprawl – the unplanned growth of urban areas into the surrounding countryside – occurs, habitats and open spaces are lost, and the green belt comes under increasing pressure to be developed. When an urban population increases, there is more traffic on the roads resulting in higher levels of air pollution, increased traffic congestion and longer journey times. Public transport may also struggle to cope with the increased number of passengers. More residents mean higher amounts of rubbish and waste to dispose of.

An overview of the distribution of population and the major cities in the UK, shows that the population distribution is uneven. Between 2011 and 2021, the number of people living in the UK increased 6%, up from 63 to 67 million. However, the rate of increase was slower than in the previous decade, due to the lowest number of births for 14 years, alongside an increase in emigration, and a fall in international immigration.

In 2021, 57 million people were living in urban areas and 10 million in rural. With 85% of the population living in towns and cities, and 15% in the countryside, this is one of the highest rates of urbanisation in the world and is a result of the process occurring over a long period of time.

Southeast and northwest England as well as the major cities of the UK are the most densely populated areas, with 5,700 people per square kilometre in London. The most sparsely populated rural areas, where the population density is fewer than 50 people per square kilometre, are located in Scotland, Wales, Northern Ireland and southwest England.

The location of raw materials such as coal and iron ore determined the location of several industrial towns in the eighteenth and nineteenth centuries. Cities that owe their origin to this, include Glasgow, Newcastle, Manchester, Sheffield and Swansea. Trade with Europe and the rest of the world led to the growth of the ports and the towns associated with them. Thus, Liverpool and Bristol became important cities. London grew because of its location on the River Thames and its close proximity to Europe but also its economic and political functions. Birmingham grew into the UK's second largest city because of trade and industry.

In the twenty-first century, UK cities are changing due to both inward and outward migration. International migrants from a variety of different countries have moved to most of the UK's cities and have formed their own communities. At the same time, many people have been moving from urban areas to seek a new life in the countryside or on the coast.

Newcastle is a case study of a major city in the UK, which illustrates its location and importance both within the UK and the wider world; the impacts of national and international migration on the growth and character of the city; and how urban change has created opportunities and challenges.

Newcastle is located in north-east England, 446 km north of London and 74 km south of the Scottish border. The city is built on the north-west bank of the River Tyne on a steep hill. This defensive site overlooks the bridge which the Great North Road from London to Scotland crosses. Newcastle has a long history of being an important settlement – stretching back to Roman times when Hadrian's Wall was built across northern England from Newcastle to the Irish Sea. The city used to be a busy port with ships transporting coal from the local coal fields to London. A large shipbuilding industry developed as a result of the coalmining and engineering industries that sprung up during the Industrial Revolution. Newcastle residents are called Geordies.

Today, Newcastle plays a key role regionally in North East England, nationally in the UK, and internationally in Europe. Newcastle is the largest city in the north-east of England and its population size has increased by 7.1%, up from around 280,200 in 2011, to 300,200 in 2021. This is higher than the overall increase for England which is 6.6%. It is part of the Tyneside conurbation, which is the eighth most populous urban area in the UK. Nationally, Newcastle is important in the UK as it is a core city with an important port and a key strategic role between the north of England and Scotland. Internationally, Newcastle is important as it a member of the Euro cities network.

Newcastle has been affected by de-industrialisation, which is the decline of traditional manufacturing industry due to the exhaustion of raw materials, loss of markets and competition from NEEs. However, Newcastle still retains some important industries, for example digital technology, offshore engineering, life sciences, creative industries and financial services. Other large employment sectors include business, retail and distribution. Some of the new industries are located near to Central Station and in the Stephenson Quarter and there are new shopping centres at Eldon Square and at the Gateshead Metrocentre.

Fifty thousand domestic and international students attend Newcastle University and Northumbria University.

Tourism is a growing economic activity in Newcastle. It is estimated that in 2018, visitors to the city spent over £1 billion attending the wide range of attractions, shops, bars and restaurants, festivals, museums, art galleries and theatres. Many also came to watch sport – Newcastle United is a Premier league football club – and to participate in events such as the annual Great North Run. The top attractions include: the bridges over the Tyne, the historic quayside, Hadrian's Wall, Grey's Monument, and the Laing Art Gallery.

Newcastle is an important transport hub with a fast rail service and a major road linking the city to London and Scotland. There is also a ferry service from North Shields to Amsterdam and an international airport. The Metro is Newcastle's urban underground railway which connects the major urban areas in Tyne and Wear and is used by over 35 million passengers every year. Network One tickets enable unlimited travel on most buses, rail and the Metro within the Tyne and Wear region.

National migration has led to the growth of Newcastle as people from other parts of the UK including Scotland, Wales and Ireland have moved there for jobs and opportunities. International migrants from Europe, the Middle East and Asia have also made Newcastle their home. Over 10% of Newcastle's population come from the Middle East and Asia. After a period of decline, Newcastle is now one of the fastest growing cities in the UK with the population growing at the rate of 0.5% each year. The city is projected to reach over 310,000 by 2030. This population growth is because of both natural and migration increase.

Migration has also changed the character of the city. The city is now more ethnically diverse. In 2011, the population was made up of 10% Asian or Asian British, 2% Black, and 86% White. The different immigrant groups have enriched the city's culture by building new places of worship such as temples, mosques and synagogues and by celebrating their festivals. There is a greater variety of food shops and restaurants. The first Chinese restaurant was opened in 1949, and now the area along Stowell Street in the historic heart of Newcastle is known as Chinatown.

Urban change has led to more opportunities for the residents. New homes, health care facilities and schools have been built for the growing population. A new primary school and a new secondary school are planned for Newbiggin Hall. Residents and visitors have benefitted from the new facilities and venues for recreation and entertainment such as the Millennium Bridge, which opened for pedestrians and cyclists in 2001.

Redevelopment is taking place in some parts of the city. For example, a new urban village is planned for Quayside West on a derelict site next to the Metro Radio Arena. It will have 1,500 homes, a hotel, bars, restaurants and several green spaces. This involves a change in land use from industrial to residential and retail, and afterwards the population density will be higher than before. The redevelopment of the city centre and the Quayside, have created new jobs in a range of sectors.

In 2018, 72% of Newcastle's population were in work, although this was below the national average of 79%. The number of people in employment is growing as several industries are creating new jobs. In 2019 Newcastle was awarded the title of "Smart City of the Year" for its innovative use of technology. The Newcastle Helix is leading development in the digital world.

The creation of new jobs and opportunities in Newcastle has led to the multiplier effect. When workers spend their money on goods and services, they pay VAT which means that the government can increase its spending on public services and infrastructure.

Newcastle used to be a dirty, noisy, industrial city with high levels of smoke and dust. Air quality improved after the closure of many factories and ship yards in the 1970s, and the introduction of electric vehicles has further reduced emissions. Over 1,000 electric car charging points are being installed in the city. The pedestrianisation of the city centre has reduced the number of cars as well.

Urban greening has led to the planting of more trees. Biodiversity has been increasing in the city's parks and open spaces. Town Moor, a large park, acts as an urban lung. The Tyne is now the best river in England for salmon and trout fishing and Newcastle has become one of the greenest cities in Britain.

Urban deprivation and housing inequalities do exist in Newcastle. For example, in Byker a south-east ward, over 73% of the people live in flats or terraced houses and 70% rent their homes, many of which are overcrowded and poorly maintained. There is also a lack of open spaces and a higher rate of crime.

In contrast, in West Gosforth, an affluent suburb of Newcastle where 50% of the residents live in semi-detached and detached houses, and only 21% live in rented accommodation, there is a lower rate of crime, more open space and better-quality housing.

Educational outcomes also vary across the city. Students from Byker leave school with lower GCSE results than those from West Gosforth, and only 16% of Byker residents have Level 4 qualifications compared to 56% of those from West Gosforth.

There are health inequalities as well. Life expectancy at birth in Newcastle for men is 75 years, which is below the national average of 77 years. People live longer and have fewer years living with a disability in the richer suburbs than in the poorer inner-city areas. Men living in West Gosforth live 13 years longer than males in Byker. Disability-free life expectancy for men and women in Newcastle is shorter than the average for England.

Newcastle's unemployment rate of 5.4% is above the national average. More than 36% of the workers are employed in the public sector which is higher than most other UK regions. The poorer parts of the city have higher levels of economic inactivity. In Byker unemployment is 8.6% compared to only 2.5% in West Gosforth. People living in West Gosforth are more likely to have professional jobs.

Economic change within Newcastle has led to several derelict sites throughout the city. Buildings are often left vacant when shops, offices and factories close. Areas earmarked for redevelopment can be left vacant for years – blighting the landscape. On some former industrial sites, the soil has been contaminated with chemical and toxic metals. The decontamination process is expensive adding to the costs of redeveloping brownfield sites. However, the site of an old leadworks near the Metro Radio Arena is currently being cleaned up so that new homes can be built.

The growing population has led to more waste. There are three recycling centres including one at Byker. Newcastle currently sends around 34,000 tonnes of rubbish to landfill each year, roughly a quarter of its total refuse. The city council is planning to use new technology to reduce this amount. This will help the city reduce its carbon emissions.

Open spaces and habitats are lost when new housing estates are built on greenfield sites, for example in South Nun's Moor and Newcastle Great Park. The countryside around Newcastle is also under threat from urban sprawl, increased traffic and air pollution as the city expands into the rural-urban fringe. The rural-urban fringe is a zone of transition between the built-up area and the countryside. In this zone there is often a mixture of land uses – some farmland and some transport, retail and recreation.

In north-west Newcastle, the expansion of commuter villages such as Ponteland is taking place in the rural-urban fringe. Improvements have been made to the local roads, leisure and social facilities in order to meet the needs of the commuters and the new residents.

However, as Newcastle grows outwards and more homes and roads are built, more green space and habitats will be lost reducing biodiversity. There is also an increased risk from flooding as more land is covered with impermeable surfaces. The increase in the number of commuters adds to traffic congestion, and air and noise pollution.

Urban regeneration is the revival of the declining parts of a town or city. Urban regeneration can be achieved by demolishing the old homes and businesses and clearing an area to make a fresh start. Alternatively, the run-down buildings can be restored and converted into new homes, shops and offices.

An example of an urban regeneration project is the Hull Fruit Market. The reasons why this area of Hull, a city on the east coast of England, needed regeneration include economic, social and environmental decline. The historic heart of Hull, which is located on the Humber estuary, was bombed during the Second World War. After the war some buildings were rebuilt, but most of the area around Humber Street became semi-derelict. Businesses closed and moved away – the old fruit market relocated in 2006. Many families left the area due to a shortage of housing and an increase in anti-social behaviour. Streets were blighted with boarded up units, poor lighting, litter, and graffiti. By the 21st century, the former commercial area near the quayside was in urgent need of urban regeneration.

The Hull Fruit Market redevelopment project began in 2013. The regeneration scheme, which cost £100 million, was led by Hull City Council and private developers. The main features of the project were the replacement of unsafe and unusable buildings with new homes, shops, business units, infrastructure and public spaces. The old warehouses and historic buildings were either restored, demolished or converted. A mixture of housing was built, with family homes and flats available to buy and rent at different prices. A new vibrant digital and creative quarter was created.

New infrastructure, including a footbridge across the A63, cycle routes, footpaths and pedestrian areas, was built to encourage economic investment, and to improve road safety and accessibility between the seaport and the historic town centre. Passengers arriving at the cruise and ferry terminals can now wander through the Fruit Market, patronize the shops and cafes and watch a live show.

The environment of the Fruit Market has been improved with landscaping, the planting of trees and the repainting of murals. The streets are kept clean, the street lighting improved and graffiti removed. Where once the Fruit Market was dirty and abandoned, now it is bustling with people of all ages attracted by the cleaner, greener, safer and brighter surroundings.

The Hull regeneration project has also built community cohesion and engagement by having a weekly street market and by opening venues which can stage live music and drama events. For example, a former dry dock, the Stage@The Dock, has been turned into an open-air theatre. The annual Freedom Festival brings families together at the Humber Street Sesh.

The positive impacts of the Hull Fruit Market project include population growth, an increase in the number of jobs and opportunities, and an increase in the number of visitors. The mixture of housing, retail, arts, crafts and business has achieved economic, social and environmental sustainability.

Urban sustainability requires the management of resources and transport. A sustainable city is one in which there is minimal damage to the environment, the economic base is sound with resources allocated fairly and jobs secure, and there is a strong sense of community, with local people involved in decisions made. Sustainable urban living includes several aims including the use of renewable resources, energy efficiency, use of public transport, accessible resources and services. Sustainable cities can provide a good quality of life for their citizens without leaving a burden on future generations. Political, economic, social, technological and environmental factors all play a part in achieving sustainability.

Political sustainability is achieved by having a city council who runs the city well. This involves collecting taxes and spending the money on ensuring that the streets are cleaned, the waste collected and there are enough schools and medical centres for everyone. It also involves employing enough police and firefighters to maintain law and order and public safety.

Economic sustainability involves ensuring that there are enough well-paid jobs and a range of opportunities for the economically active population so that they have higher incomes. This can mean attracting and keeping businesses that employ people and also running good quality public services. Social sustainability means ensuring there is enough good quality housing, education and health care for all the residents so that they can live long healthy lives.

Technological sustainability involves keeping the systems of the city running smoothly whether this is providing clean water, energy, broadband, waste collection or transport. Finally, environmental sustainability is ensuring that the city is a clean green place to live in with low levels of pollution and high levels of biodiversity.

In 2022, London was ranked sixth in the world in terms of sustainability. It has achieved this ranking because it is making good progress on social, economic and environmental indicators. The residents generally have a good quality of life with a choice of housing, education and health care. Its economy is strong with a wide choice of jobs and opportunities. London's particular strength is its environment as it has plenty of green open spaces and good air quality. The city also has a well-developed transport network with good public transport, cycle lanes and pedestrian areas. An above average number of new electric vehicles are being registered in London, even though more electric vehicle charging points are needed. London uses a lot of renewable energy but needs to do more to improve energy efficiency and reduce energy consumption.

Features of sustainable urban living include water and energy conservation, waste recycling, and creating green space. Water conservation has become an important feature of sustainable urban living as droughts and heavy downpours have become more frequent as a result of climate change. This involves a reduction in the amount of water being used and wasted, or lost to evaporation. In London, Thames Water encourages its customers to use less water by taking shorter showers, by using dishwashers and washing machines only when full, and by fitting water meters. Other ways of conserving water include fixing burst pipes and dripping taps, and fitting dual flush toilets and push taps, both of which use less water. Rainwater can also be collected by fitting gutters to sheds and other outdoor buildings and funnelling the rain into water butts.

Flooding is becoming a common problem in cities particularly after a heavy downpour. This is because the land has been covered with impervious surfaces thus preventing the rain from soaking into the ground.

Urban planners are reducing the impacts of flooding by designing and building sponge cities. These cities have more permeable surfaces such as pavements made of porous asphalt and large stones. These pavements allow rainwater to seep through them thus reducing flash flooding and costing less to de-ice in the winter. Sponge cities also have more man-made water channels and natural streams and rivers so that there is always somewhere for the surface water to flow into.

In many cities, rain gardens are being built. These are depressions in the ground filled with plants which become temporary ponds after periods of heavy rain. Bioswales are larger rain gardens. They are channels lined with clay and gravel, and planted with vegetation. They can be built on rooftops, along roads or at the side of car parks. In a storm, they become filled with water, litter, vehicle oil and other pollutants. As the dirty water flows through the bioswale, the pollutants are trapped by the plants before soaking into the ground where they decompose or are broken down by the soil bacteria. The longer the water stays in the bioswale the better the outcome.

Bioswales not only help to prevent flash flooding, they also reduce the amount of polluted water entering the city's streams and rivers, increase groundwater levels, and become wildlife habitats. In Pittsburgh where bioswales have been introduced, not only is stormwater being better managed but a more attractive cityscape has been created.

Energy conservation is another feature of sustainable urban living and involves not only using renewable forms of energy rather than burning fossil fuels but also using and wasting less energy. This benefits the environment as lower amounts of greenhouse gases are emitted.

Householders can conserve energy by switching off lights, not using standby mode, and draught-proofing windows and doors. People can also wash their clothes at lower temperatures, dry their clothes on washing lines or racks, and have short showers instead of long baths.

Motorists can reduce their energy consumption by driving electric vehicles, which cost less to run than diesel or petrol cars and have very low carbon emissions.

Using renewable energy is another effective way of conserving energy. By installing solar panels, businesses can cut their energy costs and reduce their greenhouse gas emissions. Many companies and householders have switched to light-emitting diode (LED) lighting which reduces energy consumption, fossil fuel dependency and greenhouse gas emissions.

Waste recycling is another important feature of sustainable living. Sustainable cities have well developed systems for dealing with the large amounts of waste produced by its residents every day. These cities aim to send the minimum amount of waste to landfill or incinerators, instead recycling it. Waste that can be recycled includes scrap iron and steel, aluminium cans, glass bottles, paper, wood, and plastics. Items made from finite resources such as petroleum, natural gas, coal and mineral ores, are particularly suitable for recycling, as this reduces the amount of new raw materials that are needed, and also saves on the energy required in manufacture.

When waste is recycled it is processed back into the same product or into a new item. For example, recycled paper is wound on to huge rolls before being cut and dispatched to make new products including cardboard, newsprint and office paper. It may also be used to produce items such as labels, bags and gift cards. Some 86% of recycled steel is used in furnaces and 13% is used to make new structures. Old buildings can be recycled as well by restoring and converting them or by salvaging and reusing their bricks, concrete or timber.

Creating green space is a vital feature of sustainable living. This is because green spaces and trees cool down urban areas. Urban greening is the process of increasing and preserving open space such as public parks and gardens in urban areas. Cities have higher temperatures than the surrounding countryside because the concrete and asphalt buildings and surfaces absorb the Sun's heat and reflect it back again. This is known as the heat island effect. In Tucson, Arizona more trees are being planted to increase the amount of shade and transpiration in a city which has a desert climate. Native evergreens are being planted in Seattle to soak up stormwater and to act as natural sponges.

Urban green spaces, parks and gardens have been shown to have many environmental benefits such as reducing air pollution and absorbing carbon dioxide. They also have social benefits as they improve the mental and physical health of residents. Street trees not only provide shade and create pockets of cooler air, but also help to cut air pollution by absorbing carbon dioxide and by producing oxygen during photosynthesis. Hedges also help to reduce air pollution by trapping small particulates from car exhausts. Roof gardens, balcony gardens, and living walls can all help reduce carbon emissions, as well as creating wildlife habitats.

Traffic congestion occurs when there is too great a volume of traffic for roads to cope with, so traffic jams form and traffic slows to a crawl. If there are too many vehicles on the roads of a city this results in traffic congestion, long journey times, lost productivity, lower incomes and high levels of air pollution. Carbon emissions soar increasing the greenhouse effect and adding to causes of climate change. Traffic congestion also increases levels of stress and encourages people to leave the city.

How are urban transport strategies used to reduce traffic congestion? Encouraging people to switch to alternative forms of travel such as public transport, cycling or walking can reduce congestion. This can be done by introducing integrated transport systems. These systems link bus and rail services so that when passengers arrive at their destination they can switch to a bus or train without waiting too long for the next service. This has been done in Bristol and it has resulted in a reduction in the number of cars on the road as well as reducing carbon emissions.

Another strategy is to encourage people to travel by public transport rather than by car. This can be achieved by having a single travel ticket which covers all journeys by bus, rail and underground railway in a city. The London Oyster card does this. Many cities also have bus lanes to make bus journeys shorter. If more people used public transport, this would reduce congestion, lower emissions, improve air quality, benefit health, and make cities more liveable and family friendly.

London and many other cities are promoting cycling as an alternative way to travel. Cycle lanes and bike racks have been installed in order to make cycling safer and the storage of bikes more secure. Cycling routes between town centres are signposted and maps produced to spread awareness. If more people cycled, there would be fewer cars on the roads, which would lessen congestion and speed up journey times.

Another way to discourage people from using their cars is to introduce a congestion charge. In London commuters pay to drive into the city centre during working hours. Other ways of encouraging people to leave their cars at home include incentives to car share, parking restrictions, and park and ride schemes. In Singapore, they have eased traffic congestion by introducing an Electronic Road Pricing system whereby drivers pay to drive into the city centre. This has reduced traffic congestion and the number of car accidents. Travel times are shorter which has benefitted the economy as well as helping the environment by reducing carbon emissions.

16. The changing economic world

There are global variations in economic development and quality of life. Development is the progress of a country in terms of economic growth, the use of technology and human welfare. The development gap is the difference in standards of living and wellbeing between the world's richest and poorest nations. There are some highly developed countries, where the majority of the people enjoy a good quality of life, with high levels of income, nutrition, housing, health care and education. There are also some other less developed countries, where there are lower standards of living. However, the modern world is developing at an increasing rate and is becoming more politically, economically, socially and technologically advanced.

There are different ways of classifying countries according to their level of economic development and quality of life. In the past, the Brandt line was used as a method of classification. The Brandt line ran around the world and divided countries into two groups. All the countries to the north of the line were classified as the rich north, and all the rest were deemed to be the poor south. Australia and New Zealand were included in the rich north as were the US, UK, France, Germany and Japan. Brazil, India, South Africa and Saudi Arabia, along with India and China, were classified as the poor south.

However, in the last thirty years the world has changed, and the Brandt north-south line is now out-of-date. For example, China has now become one of the world's richest nations, and Brazil and Malaysia have overtaken some European countries in terms of wealth.

Today, the World Bank classification is used instead. This classification is based on Gross National Income (GNI), which is a measurement of economic activity. GNI is calculated by dividing the total income of a country, including the value of goods and services it sells and the income earned from overseas investments, by the size of its population. This produces an average per capita figure, which is expressed in US dollars. The World Bank assigns countries to one of four groups according to their previous year's per capita GNI.

In 2022, according to the World Bank classification, the lower-income countries (LICs) had a GNI of under $1,085. The GNI of the middle-income countries (MICs) ranged from $1,086 to $13,205; and the higher-income countries (HICs) had a GNI of over $13,205.

In 2021, the world's LICs were mainly located in Africa, for example Somalia, Uganda and Rwanda. These countries are poorer than others because they have economies that are developing very slowly. Countries that had recently experienced conflict and instability such as Afghanistan also fell into this category. The country with the lowest level of economic development in the world is Burundi, a landlocked country in East Africa, which has a GNI of only $220.

Countries which have middle incomes are found in South and Central America, Asia, and Africa. Several of the countries that have middle incomes have been categorised as having newly emerging economies

(NEEs). These countries are experiencing higher rates of economic development, usually with rapid industrialisation. They differ from LICs in that they no longer rely primarily on agriculture, have made gains in infrastructure and industrial growth, and are experiencing increasing incomes and high levels of investment. Brazil, Russia, India, China and South Africa are classified as the BRICS, as they have fast-growing economies that could collectively dominate the global economy in the future; they are also NEEs.

Countries with high GNI scores have economies based on high value services. Most HICs, except for Australia and New Zealand, are located within the Northern Hemisphere. There are clusters of HICs in Western Europe, North America, the Middle East (for example Saudi Arabia, Qatar and UAE), and in East Asia including Japan, South Korea and Singapore. The country with the highest GNI – $116,000 – is Liechtenstein, a landlocked country in central Europe. The UK is a HIC with a GNI of $44,480 in 2021.

Different economic and social measures of development are used to compare countries. These include GNI, birth and death rates, infant mortality, life expectancy, people per doctor, literacy rates, access to safe water, and the Human Development Index (HDI).

Birth rates are social measures. The birth rate is the number of births in a year per 1000 of the total population. Birth rates are usually higher in countries with lower levels of economic development. For example, the birth rate in Nigeria (a MIC) in 2021 was 37 per 1000 people compared to a rate of 7 in Japan (a HIC).

Death rates are also social indicators. The death rate is the number of deaths in a year per 1000 of the total population. Death rates are rising in countries with higher levels of economic development as their populations are ageing. In 2021 Nigeria had a death rate of 11 per 1000 compared to 12.6 per 1000 in Japan.

Infant mortality, another useful social measure, is the number of deaths of infants under 1 year of age, per 1,000 live births, per year. Infant mortality rates are higher in countries with lower levels of economic development. For example, in 2021 the rate was 56 per 1000 live births in Nigeria, but only 1.9 in Japan.

Life expectancy, another social indicator, is the average number of years a person might be expected to live. As the level of economic development increases, so does life expectancy. In 2021 life expectancy in Nigeria was 55 years compared to 84 years in Japan.

The number of doctors per 1,000 people is another useful indicator of social development. As a country becomes more economically developed, the number of doctors rises. For example, in Japan in 2021 there were 1.5 doctors per 1,000 people, compared to Nigeria where there were only 0.38 doctors per 1,000.

Literacy rates, which are the percentage of people who have basic reading and writing skills is another social measure. As a country becomes richer, more people have access to education and literacy rates rise. In Nigeria, the adult literacy rate is 77% and in Japan it is 99%.

The percentage of people with access to safe water is also an important indicator. There is a clear link between the level of economic development and the access to clean water – the access rises in line with GNI. In Japan 99% of the population have access to safe water but only 21% in Nigeria do.

The Human Development Index (HDI) combines income as shown by Gross Domestic Product (GNP) per capita, which is an economic measure, with two social indicators – life expectancy and adult literacy – in order to give a more accurate measure of development and to allow comparison between countries. On the index, human development is divided into four levels. Countries with a very high level of human development have HDI scores of 0.8 to 1. For example, in 2021, the HDI of Japan was 0.92. Countries with a high human development level score 0.7 to 0.79; whereas nations with medium human development levels are 0.55 to 0.7 on the HDI scale. Countries with low human development levels score below 0.55. Nigeria measured 0.53 on the HDI in 2021.

The limitations of using economic and social measures to determine the level of development within a country have been recognised. Economic measures such as GNI can be misleading. GNI is an average figure, which does not show the extremes of wealth and poverty within a country. GNI is reported in US dollars and doesn't allow for the relative spending power of different currencies. In addition, GNI does not include the income of the people working in the informal economy or social indicators such as literacy levels.

Using birth rates as a measure of development is fairly reliable because there is a strong link between the level of development and the birth rate. As a country becomes more developed, more women stay in education for longer and have a job or career, before starting their families. This delay reduces the number of children that they have.

However, birth rates can be affected by government policies. Some countries, for example Italy, offer incentives to have larger families. The governments of other countries have tried to lower birth rates by various methods such as the One Child policy that China had for many years. In addition, birth rates may not be accurate because there may not be a reliable system of recording births, particularly in remote rural areas. Also, birth rates can vary between the different regions of a country. Lower birth rates in the cities can hide the higher rates in the countryside, thus distorting the overall picture.

Death rates are a less reliable indicator because of the ageing populations of the more developed countries and the fact that the death rate is now low in most countries. Brazil and other NEEs have younger populations which have lower death rates.

Therefore, the best way to compare countries is to look at a number of different indicators over a period of time rather than relying on one measure which may not tell the whole story.

The link between stages of the Demographic Transition Model and the level of development has also been recognised. The Demographic Transition Model is a model showing how populations should change over time in terms of their birth rates, death rates and total population size. The model shows how changes to a country's birth and death rates can result in natural increase and decrease. As a country makes the transition from low to high levels of economic development, the gap between birth and death rates at first widens and then narrows, while the total population grows and then declines.

Countries in Stage 1 of the DTM are predominantly rural with low levels of economic development. Both birth rates and death rates are high at around 36 or 37 per 1,000 people per year. During difficult years when there are famines, death rates rise as people die from food shortages and malnutrition. When there is plenty of food, the birth rates rise as women are healthier. In this stage there are low levels of GNI, life expectancy, doctors, literacy and access to safe water. The total population remains low but can be growing steadily.

In Stage 2, countries have started to develop economically, and one of the first improvements in the standard of living of people is that they have better access to a wider range of foodstuffs and cleaner water. Their health begins to improve due to their better diet and nutrition, and better access to health care such as vaccinations, medications and operations. This results in a fall in the death rate to 18 per 1,000 or even lower, as more women survive childbirth, more children live beyond their first year, and more adults live longer. The birth rate however remains high at 36 or 37 per 1,000, as it is still the tradition to have several children, in the hope that enough will survive to support their

177

parents in their old age. Nigeria is an example of a country in Stage 2. In 2021, its birth rate was 37 per 1,000 and its death rate was 11 per 1,000 and its population growth rate was 2.5% a year.

In Stage 3, the gap between birth and death rates narrows again as birth rates fall quickly down to around 17 or 18 per 1,000 people. The birth rate falls as women realise that the better health care and nutrition available has resulted in lower infant mortality rates, so there is less need to have as many children. Also, as access to education improves, many women marry and start their families later, so that they can continue studying or join the workforce. In addition, better access to contraception means that couples can control the size of their families.

Death rates in Stage 3 remain low and steady at about 15 per 1,000 people. Countries in Stage 3 experience a period of rapid population growth with high natural increase. This demographic growth results in a larger workforce and increases in GNI. As the economy grows, more jobs and opportunities are created. Families have higher incomes and increased spending power, which leads to a further increase in the economy. As tax revenues rise, governments can spend more on providing public services and infrastructure. India is an example of a country in Stage 3. In 2022, the birth rate in India was 16.82 per 1,000 people, the death rate 10.3 per 1,000, and the population growth rate was 0.67%.

In Stage 4, birth rates fall further to about 15 per 1,000, as people become wealthier and have smaller families. The death rate remains low at about 12 per 1,000 people because there are high standards of living with plentiful food, clean water, high incomes and good access to health and social care. The total population continues to grow but at a slower rate. The UK is in Stage 4. In 2022 the birth rate in the UK stood at 10.79, the death rate was 9.07, and the population growth rate was 0.53%.

A few countries are now moving into Stage 5 of the DTM. The levels of economic development are high in this stage with high GNI, high life expectancy, high literacy rates, and a high number of doctors. However, in this stage the birth rate falls to about 7 per 1,000 people, because there are more older people who are not of child bearing age. As the birth rate has fallen below the death rate, which is about 9 per 1,000, this results in a declining population as not enough babies are being born to replace the people who are dying. Countries in Stage 5 have ageing populations and a smaller workforce. In Germany in 2022 the birth rate was 9.08 per 1,000, the death rate was 11.98, and the population growth rate -0.11%.

The causes of uneven development are physical, economic and historical. The physical factors which slow down economic development include being landlocked and having a tropical climate. In addition, if a country experiences natural hazards such as earthquakes and hurricanes or weather hazards for example, floods, droughts and heatwaves this can also result in a slower rate of development.

A landlocked country is one which has no coastline and is surrounded by other countries. This situation makes trade with other countries slower and more difficult especially if a port is a long distance away. Having a tropical climate, also affects economic development as working during the heat of the day can be difficult, and the hot wet conditions result in infectious diseases such as malaria which lower the productivity of the workers. An example of a landlocked tropical country with a low level of economic development is Zimbabwe, a MIC in southern Africa with a GNI of $1,530 in 2021.

If a country experiences frequent tectonic or weather hazards, this can slow down their economic development due to the loss of life, the destruction of infrastructure, and the disruption to industrial output and trade. After the earthquake or hurricane, money has to be spent on repairing the damage and getting things back to normal which means less money is available for investment in new development projects. This has happened in Haiti, a MIC with a GNI of $1,430, where the devastation wrought by the 2010 earthquake and 2020 Hurricane Laura, has held back development by decades. Similarly, the 2022 floods in Pakistan, a MIC with a GNI of $1,470, has delayed the progress the country was making.

Economic factors also hinder development. These include having few natural resources, poor harvests, low primary product revenues, low levels of investment, high levels of debt, and high government spending.

If a country has few natural resources such as water, fertile soil and minerals this can limit economic activities like farming and mining. For example, Niger a LIC in west Africa with a GNI of $590, is landlocked with a hot dry climate, a shortage of water, few minerals, and many infectious diseases. These disadvantages mean that its economy is taking a long time to develop.

Prices for primary products fluctuate on the world market. If there is a shortage of a commodity after a poor harvest, the price increases. The country selling the crop therefore receives a higher income. However, if there is a surplus of the goods, the price falls as does the country's income. Also, transnational corporations (TNCs) buy large amounts of a product and can negotiate lower prices. This means that countries selling primary products often do not receive a reliable income. This happened to Zambia in 2015 when the world price of copper fell. Zimbabwe has also been hit by low diamond prices and poor harvests.

Cote d'Ivoire, a MIC in west Africa, is the world's largest producer of cocoa beans, the price of which fluctuates on the world market. In order to have a more reliable income, the country has started to make and export chocolate instead of exporting the primary product. This change has enabled Cote d'Ivoire to have one of the highest economic growth rates in the world.

If there is a shortage of capital and investment it can be hard to finance new factories and enterprises. If a government has a high level of debt, it will have to pay more in interest charges and will have less to spend on building new roads and other infrastructure, which could help develop the economy.

Historical factors such as colonisation can also lead to uneven development. During the colonial period, many countries in Africa were governed by European countries who shipped valuable raw materials such as timber, cocoa, gold and diamonds back to Europe. Although there were improvements to health care, education and transport infrastructure during colonisation, the economic and political development of the colonised countries was delayed.

After independence, many former colonies experienced periods of civil war and conflict. This happened in Mali, a landlocked LIC in west Africa, after it regained its independence from the French in 1960. There were years of political instability, ethnic conflict, and outbreaks of terrorism which have led to Mali being one of the poorest countries in the world, dependent on the fluctuating prices of its gold and cotton exports, and on foreign aid.

The consequences of uneven development include disparities in wealth and health, and increased international migration. Disparities in wealth occur because some of the countries with very low levels of economic development have small national incomes. A shortage of money means that the governments cannot afford to invest in developing the economy, building infrastructure or providing public services. This means that the poorest countries lack basic services such as schools and hospitals and infrastructure such as roads, resulting in lower standards of living. In contrast, the richest nations have money to invest in their economies leading to a higher quality of life.

There are also disparities in health care. In lower-income countries there are fewer doctors, nurses, midwives and hospital beds. Also, the more expensive medicines and treatments may not be available due to a lack of money to pay for them. In many poor countries there is also a higher rate of infectious diseases such as malaria and tuberculosis. A basic health service results in lower life expectancy. A baby born in Lesotho, a LIC in south Africa, will expect to live to be 51 years old. However, in HICs people live longer and die of diseases related to old age such as heart disease, cancer, and diabetes. A baby born in Japan, one of the richest countries, can expect to live to 84 years of age.

Another consequence of the uneven development of countries is a rise in international migration. People from poorer countries move to richer ones in search of higher incomes, more opportunities, a higher standard of living and a better quality of life. For example, over 16,000 people have moved from Somalia, a LIC, to the UK in recent years.

In addition, there can be uneven political development with some countries much safer and more peaceful than others. When civil war and conflict breaks out, many people choose to leave their homelands in search of safety and security for their families. Since 2011, over 6 million people have fled from the conflict in Syria with many seeking refuge in neighbouring Turkey.

An overview of the strategies which can be used to narrow the global development gap include investment, industrial development, tourism, aid, the use of intermediate technology, fairtrade, debt relief, and arranging microfinance loans. In future the difference between the standards of living and quality of life in the richest and poorest countries will be much reduced. The strategy or strategies that a country chooses to use will depend on several different factors, including its current level of development and its economic and physical geography.

Investment involves spending money on projects which enable more economic activity to take place. When new transport links such as roads and railways, ports and airports are built, people and goods can travel more freely and easily, increasing movement and trade. For example, a new 248 km rail line between Kano in northern Nigeria and Maradi in south Niger will boost trade between both countries.

Industrial development can also kickstart economic growth because opening a new factory leads to the multiplier effect. When products are manufactured new jobs and opportunities in the local area are created. This happened in Malaysia where the national car company Proton was set up in 1983 to kickstart development. The industry now makes half a million cars a year, employs 700,000 people, and contributes to Malaysia's GDP.

Tourism is another strategy which can be used by countries to develop their economies. Built attractions such as castles, Roman antiquities, cathedrals, temples and mosques can be opened to visitors. Natural attractions such as beaches, mountains, lakes, stunning scenery and exotic wildlife can also be developed for tourism. In addition, if a country has mountains and snow, a ski industry can be started, as can summer sun vacations if there are sandy beaches and beautiful coastlines.

International aid can help lower-income countries improve their standards of living. The UK gives aid to different countries in order to fund development projects. In Pakistan new teachers were trained and new schools opened. This help to educate 11 million primary school children. In Kenya aid money funded the building of 10,000 affordable new homes which are energy and water efficient. This has helped improve the health and safety of families, as well as reducing energy costs. In Tanzania, UK aid helped to deliver a £55 million family planning programme which allows women to control the size of their family, and to improve their quality of life.

Using intermediate technology is another strategy. Intermediate technology is the simple, easily learned and maintained technology used in a range of economic activities serving local needs in LICs. This means using tools and machines which are simpler to operate and do not run on petrol or diesel, require servicing or spare parts. For example, in rural south east Nigeria, bicycles are used by farmers to take their crops to market instead of using a truck. Taxi drivers use bicycles rather than a car to transport their passengers. Not only are bicycles a sustainable form of transport, but they are more reliable and cheaper than cars and can be used to travel to inaccessible parts of the countryside.

Fairtrade is when producers in LICs are given a better price for the goods they produce. Often this is from farm products like cocoa, coffee or cotton. The better price improves income and reduces exploitation. Fairtrade is also a system of certification that aims to ensure a set of standards are met in the production and supply of a product or ingredient. Farmers receive a premium price for their Fairtrade products. For farm workers, Fairtrade means that they gain workers rights, safer working conditions and fairer pay. For shoppers, the Fairtrade logo indicates good quality and ethically produced products. In Ghana, most bananas produced are Fairtrade. This has raised income and standards of living. The extra money has funded community development projects such as improving the local school, paying for secondary education, providing free healthcare, and distributing mosquito nets to protect against malaria.

Debt relief is another strategy which can help reduce the global development gap. Highly indebted poor countries are nations which borrowed money to pay for development projects, but then struggled to repay their loans. Banks such as the International Monetary Fund can offer them debt relief which cancels or reduces their debts, or make the repayments more manageable. Cote d'Ivoire has used debt relief to reduce the $8 billion it owes to creditors. By reducing its interest payments, the country has more money available to spend on education and health care. This helps improve the standard of living of the people and grow the economy as the workers will be healthier, better educated and live longer.

Microfinance loans are small amounts of money which are lent to entrepreneurs in the LICs to start small businesses. For example, subsistence farmers switching to cash cropping take out microfinance loans to buy seeds and equipment. In Nigeria, there is a microfinance bank called ASHA, which gives small loans to women wishing to start or grow a business.

Kenya is an example of how the growth of tourism in an LIC or NEE helps to reduce the development gap. Kenya, in eastern Africa, is a MIC with a GNI of $2,080 in 2021. The country has a beautiful coastline, a tropical climate, and abundant and varied wildlife on the plains of the interior. Over a million tourists visit Kenya every year. Visitors can fly to Nairobi and combine a beach holiday on the coast near Mombasa, with a safari viewing animals such as lions, elephants and giraffes in the Maasai Mara.

In recent years, over 600,000 jobs have been created in Kenya in hospitality and catering, and travel and tourism. It is estimated that 12% of Kenya's workforce is employed in the tourist industry either directly or indirectly. Tourism also brings in dollars and other foreign currencies, which have a better buying power than the Kenyan shilling. Kenya's tourism industry earned $2.13 billion in 2022.

The growth of the tourism industry has led to the multiplier effect. The jobs created by tourism mean that many people earn a regular wage, which they can use to buy goods and services and pay taxes. This extra money in the economy, generates further jobs and opportunities. The increased government revenues pay for improvements to the country's infrastructure. For example, Kenya now has four international airports, a road and railway network, a deep-sea port at Mombasa, an expanding energy sector, and digital telecommunications. There have also been improvements in public services such as health and education. This has led to rises in life expectancy and literacy rates, as well as a healthier, better educated, more productive workforce.

However, the tourist industry does have some negative effects. Firstly, if the hotel chains and tour operators are foreign owned, the profits can be banked in another country. This is an example of economic leakage – where the money that is spent by a tourist in Kenya does not stay and spread throughout the country, but instead leaks out to another economy. Secondly, if tourists stay in any of the all-inclusive resorts, they may not spend any money outside the resort, thereby not benefitting any of the local businesses. Overall, though, the growth of tourism has enabled Kenya's economy to develop faster than it would have done without it.

Some LICs and NEEs are experiencing rapid economic development which leads to significant social, environmental and cultural change. When economic growth occurs in a country this also results in other changes. For instance, because new industries are opening up there are more permanent full-time well-paid jobs available. This means that families have more disposable income and can improve their homes, clothes and the type of transport they use. Due to economic development leading to more urbanisation, many more people live in cities than before. An urban lifestyle is different from a rural one in that there are more opportunities to experience different types of food, meet new people and enjoy leisure and recreation facilities. The environment also undergoes a change from being wilder to one that is more influenced by human activity.

Nigeria is a case study of a newly emerging economy which is experiencing rapid economic development and significant social, environmental and cultural change.

What is the location and importance of Nigeria, both regionally and globally? Nigeria is located in west Africa. The country borders the Gulf of Guinea and lies between Benin and Cameroon. Nigeria is both a regionally and globally important country, because it has a large and growing population and economy. Nigeria is a newly emerging economy with a GNI of $2,080 in 2021. However, there are large inequalities in wealth with over 62% of the people living in extreme poverty, in stark contrast to the thousands of billionaires. There are many other inequalities as well – in housing provision, well-being, employment, access to services, open land, and safety and security. Sometimes called the Giant of Africa, Nigeria remains near the bottom of the Human Development Index. Corruption remains a major problem.

What is the wider political, social, cultural and environmental context within which Nigeria is placed? The wider political context of Nigeria is that the country is a former colony of Britain. After gaining its independence in 1960 and becoming a presidential republic, Nigeria has developed its links regionally and globally. The country is a key member of the African Union and has joined the Economic Community of West African States (ECOWAS). It is also a member of several international organisations including the Commonwealth and the United Nations, for which it has a peacekeeping role. As a major oil producing nation, it is an important member of the Organisation of Petroleum Exporting Countries (OPEC).

The social context of Nigeria is that in 2022 the population stood at 225 million. There is a 2.5% rate of population growth, which means that by 2050 the population could be 400 million, making Nigeria the world's fourth largest country. The country is in Stage 2 of the DTM. Birth rates are high and there is a youthful population. Death rates are falling although infant mortality remains high at 57 deaths per 1000, and life expectancy is only 61 years.

Culturally, Nigeria is made up of many different ethnic groups including the Hausa in the north, and the Yoruba and Igbo in the south. Although the majority of people are Muslim, there are large groups of Christians and other religions. This makes for a rich cultural scene with many different influences although ethnic tensions arise from time to time. While largely peaceful today, the country has experienced periods of political instability with outbreaks of ethnic violence. However, Nigerian music, films and literature are of increasing global importance.

Nigeria is a large country with a variety of environments and climates. The south has an equatorial climate; the centre a tropical climate and in the north, there is a desert climate. Although nearly 10% of the country is still covered with rainforest, much of the land has been cleared for farming. There are many environmental problems including air and water pollution, deforestation, soil degradation, loss of arable land, and oil pollution.

How is the industrial structure of Nigeria changing? Industrial structure is the relative proportion of the workforce employed in different sectors of the economy (primary, secondary, tertiary and quaternary). The industrial structure of Nigeria is changing as the country develops. In 1999, over 70% of the economically active workforce was working in primary industry, 10% in the secondary sector and 20% in tertiary. By 2022, the proportion of people working in the primary sector, which is made up of agriculture, forestry and mining, had fallen due to mechanization and modernisation, although the oil industry remains very important.

What is the balance between different sectors of the Nigerian economy? More people now work in the secondary sector in activities concerned with making goods from raw materials for example, food processing, steel making and car assembly. An increasing number of Nigerians are working in the tertiary sector in economic activities that provide various services – such as retail, banking, health and social care, education, entertainment and personal services such as hairdressers and fitness trainers.

How can manufacturing industry stimulate economic development? The growth of manufacturing industry in Nigeria has helped stimulate further economic development through the multiplier effect. When a factory opens, both the local and the national economy grows. The workers earn a disposable income which can be spent on goods and services, the domestic market expands creating a consumer society, which in turn creates other jobs. The government also receives higher tax revenues both from the workers and businesses, which can be spent on improving transport infrastructure, education and healthcare resulting in a healthier, better educated and higher paid workforce. This in turn stimulates industrial growth in other sectors and increases national wealth.

For example, in food processing factories, agricultural crops can be turned into a product which adds value and has a longer shelf life than the fresh item. There are now factories in Nigeria making a range of products including chocolate, instant coffee, sauces, ice creams and stock cubes. These factories employ workers in a range of different jobs including machine operators, technicians, accountants and cleaners.

What is the role of transnational corporations (TNCs) in relation to industrial development? A transnational corporation (TNC) is a large company that has operations (factories, offices, research and development, shops) in more than one country. Many TNCs are large and have well-known brands. Usually, their headquarters are in the country where their company first started. As the corporation grows they expand into other countries and open new factories, offices, shops or restaurants. TNCs play an important role in the industrial development of low-income countries because they can transfer their money, staff, knowledge and expertise to the new location.

An example of a TNC operating in Nigeria is Shell, an international energy company. Shell, one of the first companies to find oil and natural gas in the Niger Delta, started exporting oil from Nigeria in 1958. Today its subsidiary the Shell Petroleum Development Company of Nigeria (SPDC), produces an average of 739,000 barrels of oil per day which is 39% of the country's oil. The TNC is focused on onshore and shallow water oil and gas production and has 6,000 kms of pipelines, 8 gas plants, 1,000 producing wells and a deepwater field at Bonga.

What are the advantages and disadvantages of a TNC to Nigeria as the host country? The advantages that a TNC like Shell brings to a host country like Nigeria are mainly economic – investment, employment and tax revenues. Shell has invested large amounts in developing the oil industry in Nigeria. The company has funded the construction of oil wells, pipelines and oil refineries in order to produce, refine and transport oil and gas to its domestic and overseas customers.

The development of the oil industry has brought many jobs to Nigeria. Shell employs more than 4,500 people directly of whom 95% are Nigerians. Some 66% of the Nigerian staff members are from the Niger Delta. The opportunities that the industry has provided have led to higher wages and a more highly skilled workforce.

Another 20,000 people are employed indirectly through the network of companies that provide supplies and services to Shell. The corporation awards over 90% of its contracts to Nigerian companies stimulating growth in other sectors of the economy.

Shell also pays a large amount of tax to the government. The oil and gas industry accounts for 90% of Nigeria's export income and 75% of government revenue. Shell also funds development projects in the Niger Delta which aim to improve the health and education of local communities.

Hosting a TNC like Shell can bring disadvantages to the host country, including pollution, social unrest, and exploitation. Pollution caused as a result of industrial development is a major concern. In Nigeria the main disadvantage of having a large oil and gas industry has been the resulting environmental damage. Oil spills and oil flares have been a frequent occurrence in the Niger Delta. These are caused by old pipelines perishing and leaking but also by acts of sabotage, theft and illegal refining carried out by criminals.

When farmland and rivers become contaminated with oil and other pollutants, crop yields fall, ground water supplies become polluted and fewer fish are caught. This leads to a fall in the incomes of farmers and fishers, and an increase in the number of deaths and illnesses in the local people. The Niger Delta, already a poor region, has been further impoverished by this environmental damage.

The amount of money that the oil industry has brought to Nigeria can also be problematic. Oil wealth has provoked armed conflict and new militant groups such as the Niger Delta Avengers who have attacked oil-producing facilities causing their shutdown and the loss of government revenue.

Law suits can also be brought against TNCs as local people seek compensation for the loss of their livelihoods caused by oil pollution. After being found to be responsible for the oil spills and multiple leaks from oil pipelines in the Niger Delta, SPDC has been ordered to pay £13 million in compensation to local farmers. If the host country becomes too hostile an environment for a TNC to operate in, the TNC can suddenly relocate leading to a loss of jobs and the closure of factories.

Another problem can be the exploitation of workers who may be paid very low amounts to work for very long hours. Workers in a low-income country may not have the same protections that their counterparts in a high-income country have.

In conclusion, it is fair to say that overall, the presence of a TNC such as Shell has benefitted Nigeria, the host country, economically, as they have provided the resources and expertise to develop the oil industry. Going forward however, any future industrial development has to be environmentally sustainable. Most TNCs are recognising this fact.

How are Nigeria's political and trading relationships with the wider world changing? Since gaining independence, Nigeria has retained close cultural and social links with Britain and is an important member of the Commonwealth. However, Nigeria has been forging stronger political ties not only with other African nations, but also with countries such as China and India, who are becoming increasingly important as trading partners and sources of foreign direct investment. In recent decades, China has invested a lot of money in Nigeria, and is currently paying for the development of a new oilfield and the construction of a new railway along the coast. This is said to be a type of economic colonialism.

Nigeria also has changing trading relationships with the wider world. In the past, the country traded mainly with Britain exporting primary products and importing manufactured goods. Since independence, Nigeria has started trading with a wider range of countries including China, India, the EU, and the US. The UK now accounts for only $1.2 billion of Nigeria's trade compared with $15 billion with the US and India combined.

However, primary products such as crude petroleum, natural gas, beef and cocoa beans still make up a large proportion of Nigeria's exports. The majority of its imports comprise manufactured goods including refined petroleum, cars, wheat, laboratory glassware, and packaged medicines.

What are the types of international aid? International aid is the money, goods and services given by the government of one country or a multilateral institution such as the World Bank or International Monetary Fund to help the quality of life and economy of another country. Short-term humanitarian aid is given to save lives in the event of a natural disaster. Long-term developmental aid is provided to help economic development and improve standards of living. In spite of having a growing economy, Nigeria still relies on large amounts of international aid from high-income countries including the UK.

What are the impacts of aid on Nigeria, the receiving country? In 2022, Britain provided £15 million of emergency humanitarian aid to north-east Nigeria. The lives of over 8.4 million people were at risk because of extreme hunger and malnutrition caused by conflict and climate change. Working with its partners the Red Cross, the World Food Programme and UNICEF, the UK government provided emergency shelter, food assistance and nutrition. The aid agencies also improved sanitation by providing clean water and disposing of waste. Both these measures protected the health of the people, and particularly the women and children, worst hit by the crisis.

The emergency funding also supported the Nigerian government's efforts to bring peace to the north of the country by reducing violence, displacement, poverty and climate shocks. The short-term impacts of this aid on Nigeria, the receiving country, included saving lives and livelihoods.

Nigeria also receives long-term development funding for various projects which aim to improve the lives of the 100 million Nigerians living on less than £1 a day. In 2016, the UK government sent £305 million of aid to help the country provide health care, education and shelter for many people living in poverty. The money funded educational projects, which increased the number of school places for girls in a country where 8.5 million children do not go to school. Educating girls improves their life chances by enabling them to become better educated and more able to earn higher wages in a more skilled job.

Other projects focus on health care. More than 100 women die every day in Nigeria from complications during pregnancy and childbirth. To reduce this death rate, access to contraceptives and family planning has been improved resulting in fewer maternal deaths from unsafe abortions. Infant mortality rates have also been reduced by improving antenatal and postnatal care in rural areas, by treating preventable diseases, and by tackling malnutrition. Stunting in children has been prevented by improving the diets of at-risk children.

Other long-term aid projects work with the Nigerian government to improve local and national government, the justice system and the security services. These measures will help improve the running of the country and make it safer.

The economy of Nigeria has benefitted from several long-term aid projects which have funded improvements to the infrastructure, power supplies and roads. In addition, farmers markets have been helped to become established in the Niger Delta. These measures have facilitated economic activity.

Overall, the long-term impact of international aid will be that the people of Nigeria will have higher standards of living and a better quality of life.

What are the environmental impacts of economic development? The downside of economic development is that there are many negative impacts on the environment. In Nigerian cities air pollution is a major problem. The number of cars has risen with the increase in disposable incomes. Many of the older vehicles run on diesel and leaded petrol, and belch out exhaust fumes containing carbon dioxide and particulates. Other causes of air pollution are factory emissions, wood-burning stoves, and using oil and gas to generate energy. These all add pollutants, smoke and dust to the mix.

The polluted air results in regular smogs in Lagos and the other big cities. These smogs not only reduce visibility but are a health hazard. Smogs contain high levels of sulphur dioxide and nitrogen oxide which can cause a range of illnesses including headaches, nausea and respiratory diseases such as asthma and emphysema. When pollution levels are high, people are advised to stay indoors resulting in children missing school, families losing income and businesses having fewer customers.

Water pollution is another environmental impact of economic development. Water pollution is caused by waste from households and industries. Many communities lack sewage treatment plants, so untreated sewage is discharged into streams and rivers, leading to the build-up of bacteria such as e-coli. There are few regulations regarding industrial pollution and those that there are, are poorly enforced. Therefore, some factories discharge a range of pollutants, including heavy metals and oil, directly into nearby streams and rivers.

When supplies of water from wells and rivers become polluted, people may have to travel further to fetch clean water. These longer journeys mean less time for education or for earning a livelihood. If people drink the contaminated water, they run the risk of catching diseases such as cholera, diarrhoea, dysentery, hepatitis A, typhoid and polio. Incomes can be reduced if people cannot work or have to pay for health care.

A lot of household waste in Nigeria is sent to landfill, where lead, other heavy metals, and the toxic chemicals found in e-waste, seep into the ground water or wash into lakes and rivers. Illegal fly-tipping also adds to the problem. Contaminated irrigation water and soils result in lower crop yields, food shortages and an increase in childhood malnutrition.

Wildlife is adversely affected by water pollution. The discharge of a large amount of sewage into a river can overwhelm the ecosystem resulting in the death of fish. Contamination by heavy metals such as cadmium and mercury, can also have long term effects on fish and shellfish. Eutrophication can occur if nutrients from fertilizers are washed from the fields into the lakes or rivers. The consequent dense growth of plant life can suffocate fish.

What are the effects of economic development on quality of life for the population of Nigeria? In spite of the environmental damage that is occurring, economic development is having a positive effect on the standards of living and the quality of life of most Nigerians. Life expectancy has improved and is now 61 years, which is 16 years higher than it was in 1980. This improvement is due to better nutrition, housing, education and health care all of which has led to a fall in the death rate particularly for children under five.

An analysis of the Human Development Index (HDI) shows that Nigeria is making progress in terms of improving life expectancy, adult literacy and economic growth. Nigeria's HDI has risen from 0.22 in 1980 to 0.53 in 2021. This increase is one of the largest in the world and is the result of the growing economy and the multiplier effect. The increase in the number of better paid jobs and in disposable incomes has meant that people can afford to spend money on consumer goods and services, such as clothing, food and recreation, all of which improve their quality of life.

However, even though there has been a general improvement in the standard of living and the quality of life of ordinary Nigerians, some people have been negatively impacted by economic development. Both air and water pollution can cause illness which results in lost income and higher health care costs. Living with smog and dirty water also lowers the general quality of life and reduces wellbeing. Factory workers can also be adversely affected by low standards of health and safety, which can result in work-related injuries and illness, leading to a loss of income.

Major changes in the economy of the UK have affected, and will continue to affect, employment patterns and regional growth. In recent decades, the economy and the industrial structure of the UK have had to respond to the decline of major industries such as coalmining and shipbuilding.

The industrial structure of the UK is made up of four different sectors: primary, secondary, tertiary and quaternary. Primary industry is concerned with the production of raw materials such as crops, timber and fish, and the extraction from the ground of other primary products for example stone, mineral ores and fuels. Agriculture, forestry, fishing, mining and quarrying are all types of primary industry.

Secondary industry manufactures, produces and builds goods and products from a range of different raw materials. Heavy industry occurs in large scale plants, involves complex machinery, requires skilled workers, and produces products on a large scale. Examples include oil refining, iron and steel making, motor vehicle manufacture, cement production, and hydroelectric power generation. Construction and a range of light industries such as the manufacture of clothing, processed foods and computers, are included within the secondary sector.

Tertiary industry provides a wide range of services including finance, retail, tourism, hotels, health and social care, the police, and defence. Quaternary industry is involved with information and knowledge and includes industries such as information technology, research and development, the media, and education.

In the UK, primary and secondary industries are declining in importance, whilst the tertiary and quaternary sectors are continuing to grow. This has affected employment patterns. In 1991, 3% of the work force was employed in the primary sector, but by 2022 this had fallen to 1%. The proportion of people working in secondary industry fell from 28% in 1991, to 13% in 2021. Of these, 2.6 million jobs, or 7% of the total, were in manufacturing industries and 6% or 2.3 million jobs were in construction. During the same period, the number of jobs in the tertiary and quaternary sectors grew. By 2021, 84% of the UK workforce was employed in the sectors, which accounted for 30 million jobs and 80% of the economy.

The changes in the UK economy have also impacted on regional and urban growth. The expansion of the tertiary and quaternary sectors has mainly occurred in the south and east of England, where several of the top ten fastest growing cities are located. These include Milton Keynes, Peterborough, Reading, Oxford, Brighton, Inner London, Southampton and Swindon. These cities are the home to several fast-growing sectors and have benefitted from overseas investment.

Although the East Midlands are still important for manufacturing, the West Midlands have a faster rate of economic growth, and Birmingham and Coventry are listed amongst the ten fastest growing cities in the UK.

In Wales, employment is growing in Cardiff and manufacturing remains strong, particularly in the south of the country. In Scotland, Edinburgh is one of the fastest growing UK cities.

However, over half of the slowest growing economies and cities are located in the north of England. These include York and Rotherham.

The slower growing cities are all home to declining industries, such as fishing in Grimsby, and shipbuilding in Plymouth. Some cities have both a falling population and fewer jobs. These include Penzance, Aberystwyth and Whitby, which are all inaccessible coastal towns.

The economic future of the UK is one of more change. What are the causes of economic change? Economic change in the UK. has occurred because of the de-industrialisation of the economy, the decline of the traditional industrial base, globalisation and government policies.

De-industrialisation is the decline of a country's traditional manufacturing industry due to exhaustion of raw materials, loss of markets and competition from NEEs. Heavy industries such as coal mining, shipbuilding and steel manufacture were once the backbone of the British economy employing millions of people and generating large amounts of money.

The coal industry reached its peak in the early part of the nineteenth century when over a million people were employed in the coal mines. The decline in the industry after the First World War started as other sources of fuel like oil became available, and cheaper coal from other countries, such as Russia and the US came onto the market. The decline continued and by the end of the Second World War the number of jobs in the coal industry had fallen to 800,000. As there was a switch from coal to natural gas for electricity generation, the consumption of coal fell from a peak of 157 million tonnes a year in 1970, to 18 million tonnes in 2016.

In the 1970s and 1980s the British shipbuilding industry suffered a similar decline as a result of competition from countries such as South Korea. However, the shipbuilding industry currently supports 42,600 jobs across the UK and contributed £2.8 billion to the economy in 2020. The UK government is backing a revival of the industry.

The decline of the traditional industrial base began in the twentieth century, when many of the long-established manufacturing industries that had once powered the UK economy started to lose customers. The coalfields, which had been so important, became exhausted when the remaining coal became too difficult and expensive to mine. The steel industry became uneconomic as many steel works, for example in Sheffield, were located inland in high-cost locations far from a port. The textile industry, which had relied on plentiful supplies of imported cotton, also became uncompetitive when cheaper imports of cloth came onto the market.

Older industries such as steel making and shipbuilding were replaced by new industries producing high-value products, such as aeroplanes and electronic goods. These industries adapted to the new economic world by introducing modern technology, mechanisation and robots, which enabled them to operate with fewer workers and to become more efficient.

Globalisation is a process whereby the world becomes better connected through the free movement of goods, services, information and people. Globalisation has led to the UK economy changing. As new economies such as China, Malaysia and Indonesia emerged they could manufacture and sell their products more cheaply than the UK because their costs were lower. Many consumer goods such as shoes, mobile phones and televisions became cheaper to import than to be made in the UK. This led to many UK businesses and factories being unable to compete and having to close.

UK businesses started to specialise in high-end products such as racing cars, or moved into financial and business services making use of the developments in computers. This led to the growth of the financial sector based in the City of London which provides banking and insurance services on a worldwide scale.

Government policies have also shaped and supported the UK economy. An important government policy is the level of taxation, which can be used to grow or slow the economy. For example, if Value Added Tax (VAT) is reduced, this results in lower prices for goods and services, which increases consumer demand. The government can also use procurement to stimulate an industry. For example, when a new nuclear submarine is ordered this stimulates economic activity in Barrow-in-Furness where the Royal Navy's submarines are made. A major new railway line such as High Speed Two (HS2) also leads to new jobs and opportunities for many workers and businesses.

Government policies such as the goal to achieve net zero carbon emissions can also influence industry and bring about change. For example, new petrol and diesel cars and vans are set to be banned from sale in the UK in 2035. This will encourage car manufacturers to switch to making electric vehicles. The move away from using fossil fuels towards renewable energy is also bringing about change in the energy industry.

A post-industrial economy is an economy where most of the employment is in service industries. As the UK moves towards a post-industrial economy and a new economic future, several changes are taking place. Information technology (IT) is now an important part of the economy. Information technologies use computers, the internet, and mobile phone and satellite technologies to speed up communication and the flow of information. Information technology and container transport have enabled countries, people and businesses to trade and to communicate more easily with each other. In 2021, two million workers, 6.3% of the workforce, were in IT related jobs such as cyber security, game development, and web design. The sector is dominated by large companies including Amazon, Google and IBM.

Another change has been the continuing growth of the service sector Service industries, which are also known as tertiary industries, are economic activities that provide various services. These include commercial services such as retail and banking, professional services like solicitors and dentists, education, health and social care, entertainmen and hospitality, and personal services like hairdressing. In the post industrial economy, the majority of workers are now employed in the tertiary and quaternary sectors. Retail is the largest employer employing 4.4 million people in the UK. The move to working from home (WFH) which has been enabled by the extension of the broadband network, is also having an impact on employment patterns and regional growth.

Financial services have grown in importance and in 2021 contributed £174 billion to the UK economy, 8% of the total output The sector is dominated by large companies such as KPMG, Barclays and HSBC providing employment opportunities in accounting, banking and insurance.

The UK is internationally recognised for its leadership in research and development (R&D) and the excellence of its scientific institutions. In 2019, 500,000 people worked in R&D which is often based in university towns and cities, where there are a higher number of graduates. The pharmaceutical company Glaxo Smith Klein, invested £4.5 billion in R&D in 2019. British researchers were at the forefront of global effort to find a vaccine for Covid-19 during the 2020 pandemic.

A lot of high-tech industry is now sited on science and business parks Science parks are purpose built industrial estates where scientific research and commercial development are carried out. Business park are also purpose-built areas of offices, industrial units and warehouses These parks are often located at the edge of a city, on a main road and near to the local university. This is to enable workers, raw materials and finished goods to have easy access to transport links, the latest academic research, and facilities for shopping and recreation. Similar enterprise often cluster together so that they can benefit from each other's expertise and knowledge. in co-operation with the local university.

Cobalt Business Park in Newcastle-upon-Tyne is one of the largest in the country. The business park has good transport links with the rest of north-east England and is home to several large companies including Sage and Leeds Building Society. Cambridge Science Park in east England, has many biotech and technology businesses which have strong links with the University of Cambridge.

The impacts that industry makes on the physical environment include various types of pollution, soil contamination, landscape damage and habitat destruction. Factory emissions and smells, smoke, dust and vehicle exhaust fumes can all pollute the air. The discharge of toxic chemicals, raw sewage and oil into streams and rivers causes water pollution. Noise pollution can come from the operation of heavy machinery and the transport of materials. Soils can be contaminated by industrial processes and by leakage from landfill sites. Views and landscapes can be spoilt by the building of factories or by activities such as open cast mining which leave open scars on the countryside. Wildlife habitats can be destroyed when land is cleared for a new industry.

However, in recent years de-industrialisation has had a positive effect on the environment. When factories, mines and quarries close, their buildings can be demolished, the land decontaminated and reclaimed for new land uses including housing and nature reserves.

Environmental sustainability is about meeting the needs of the present generation without harming the ability of future generations to meet their own needs. It is about allowing industry and businesses to operate profitably without damaging the environment. To become more sustainable industry can switch to cleaner technology, use renewable resources and energy, reuse and recycle non-renewable resources and reduce the amount of waste they produce.

An example of a modern industrial development that is becoming more environmentally sustainable is Bakkavor, a food processing company making fresh prepared food in Park Royal, west London. The company makes ready meals from ingredients and materials which are bought from sustainable sources here and abroad. At present they use plastic food packaging because it is light, cheap and good at extending product shelf life but in the future, they hope to use sustainable packaging that is 100% reusable, recyclable or compostable.

Bakkavor reuses and recycles their non-renewable resources and uses renewable energy. They have reduced their carbon emissions by encouraging their employees to travel to work by public transport, on foot or by bike instead of by car.

Another example of sustainable industry can be found in Leicester where the Walkers crisp factory produces 11 million bags of the potato snack every day. Parent company PepsiCo has set a target of net zero greenhouse gas emissions by 2040. In order to achieve this reduction, the company is investing over £58 million in new technology, such as replacing the gas-fired ovens with new electric ones powered by renewable energy.

Social and economic changes have also been taking place in the rural landscapes of the UK. South Norfolk is a rural county in east England. It has a growing population attracted by its jobs, accessibility, lowland, favourable climate and good soils. Between 2011 and 2021 the population of South Norfolk increased 14%, a higher increase than the 7% experienced by the rest of England. This growth in population is occurring in all age groups, including a 10% jump in the number of working age adults and 12% more children aged under 15 years.

The increase in the number of children has led to a higher demand for education and leisure services. However, there has also been a 30%

growth in the number of older people aged 65 years and over. This means more demand for health and social care services.

An increase in the number of people in South Norfolk, has led to economic growth and the multiplier effect. There are more workers, who are earning higher wages and paying taxes. This has led to increased spending in the local shops and higher council tax revenues. Local businesses have benefitted from having more customers, but the demand for homes is pushing up house prices. There is more traffic congestion on the roads because of the increase in the number of commuters.

In contrast, Copeland a rural county in north-west England, has a declining population due to its highland, poor soils, remoteness and lack of jobs. Between 2011 and 2021, the overall population fell to 67,100. This was a decrease of 5%, with a decrease of 11% in the number of adults aged 15 to 64 years, and a 7.5% fall in the number of children aged 15 and under. However, the number of people aged 65 and over increased by 17%.

The ageing population and the outward migration of working people has led to economic decline. Village schools and shops have closed as families move away leaving the elderly behind. Other services have become economically unviable and have been forced to close. Many rural communities are becoming socially unsustainable as the population falls. Many rural buildings and homes are left empty and in a state of disrepair.

Improvements and new developments in road and rail infrastructure, port and airport capacity are taking place across the UK. Slow journey times, traffic congestion and poor connectivity increase costs and reduce growth. To solve these problems and to stimulate economic growth, the government is spending almost £200 billion each year improving the UK's transport network.

The road network is being developed by building new roads, widening existing ones and upgrading motorways. Recently the plan to introduce more smart motorways, which use the hard shoulder as a running lane and have variable speed limits, have been curtailed. Other ways of controlling the flow of traffic, increasing capacity, reducing congestion and improving safety will be used instead.

In addition, there are some major infrastructure projects being built. These include the Lower Thames Crossing which is a 2.6 miles long tunnel under the River Thames. As well as the tunnel, the project involves building 14.3 miles of new road to connect the M2, A13 and M25 doubling the road capacity of the road network east of London. The tunnel when it opens in 2030 will increase investment and business opportunities not only in the south-east but in the rest of the country as well.

Rail travel is also being continuously improved in order to connect more places and to make journey times shorter. Crossrail cost £18 billion and opened as the Elizabeth Line in May 2022. This new rail link enables people to travel right across London from Reading and Heathrow in the west to Shenfield and Abbey Wood in the east, via new tunnels 21km long under central London. This rail upgrade has increased London's rail capacity, cut passenger journey times, encouraged regeneration and generated many jobs.

HS2 is a brand-new high-speed railway line designed to carry over 300,000 passengers a day, travelling at up to 225 mph, on state-of-the-art zero carbon trains. The first services will connect London and Birmingham and will start between 2029 and 2033. Phase 2, which was cancelled in 2023, would have connected the West Midlands with the North.

Old Oak Common in west London is a new transport super-hub set to be the best-connected and largest new railway station ever built in the UK. The station will have high-speed rail services to the Midlands,

Scotland and the North, and access to central London and Heathrow via the Elizabeth line.

HS2 will add extra capacity to the rail network by moving long-distance services onto their own tracks. This will result in more rush hour trains and seats for passengers as well as fewer delays and reduced overcrowding on other suburban trains.

The new railway will connect eight of the UK's ten largest cities including London, Birmingham and Edinburgh and will cut journey times. The faster trains and new stations will make it easier for people to travel to other parts of the country for business and leisure. This should create more jobs in the travel and tourism industries and stimulate economic growth in the Midlands and the North.

The UK has several major ports. Felixstowe, a deepwater port on the east coast, is Britain's biggest and busiest container port, and one of the largest in Europe. It can handle the largest container ships which are a key component in handling the UK's trade.

The government is not planning to build any new ports although private companies are investing in the upgrade of existing ones. For example, Liverpool 2 docks have been expanded to become the UK's largest transatlantic port and one of Europe's most advanced container terminals. Liverpool 2's increased capacity has created many new jobs, has helped boost the UK's export earnings and has stimulated economic growth in the region.

Air travel is growing in importance. Several of the UK's major airports are expanding their capacities. Southampton Airport is extending its runway by 164 metres so that larger planes can land and take off. A new flight from London Stansted to Bergen will offer travellers from London and the East of England easy access to Norway's fjords from March 2023.

Heathrow Airport is the UK's largest, busiest and most important airport. In 2019, goods worth £188 billion and 19 million passengers passed through the airport. Many of the goods were high-value ones such as microchips which were exported by industries located in the Midlands, Wales and the North East. It is estimated that Heathrow Airport contributes £5 billion to the UK economy and supports more than 140,000 jobs.

Heathrow has plans to expand by building a third runway. This will increase the airport's capacity and will allow London to maintain its position as one of the world's leading financial centres. The larger airport will cost £18.6 billion but will enable flights to forty new destinations and create 70,000 new jobs by 2050. The proposal has largely been backed by business groups. However, considerable opposition has come from local communities and climate campaigners who are concerned about the impacts on noise and air quality in the surrounding areas.

However, all the planned improvements to the UK's transport infrastructure will boost growth regionally and nationally. Better connectivity enables people and goods to travel more easily around the country opening up new employment and leisure opportunities for millions of people.

The north-south divide is the division of the UK into two parts based on differences in culture, wealth, health, education and political influence. The south includes London, the South East, the South West, and the East of England. While the north comprises the North East, the North West, Yorkshire and the Humber, the East Midlands, the West Midlands, Wales, Scotland and Northern Ireland.

A lot of the north-south divide can be explained by the success of the City of London. The City of London provides financial and professional services which in 2022 produced £278 billion in economic output, contributed nearly £100 billion in taxes, exported over £128 billion and generated nearly £2 billion in foreign direct investment for the UK.

economy. The City drives the UK economy, generating over £85 billion in economic output annually. In 2023 there were 587,000 workers in the City of London – many of whom are young and highly skilled. City jobs have grown over 8% since pre-pandemic 2019 to 2021, with nearly 45,000 more jobs than in 2019.

In general, the south of the UK is richer, healthier and better educated than the north. People living in the south have higher standards of living and a better quality of life. The pattern of regional economic growth shows that both Wales and Northern Ireland are growing more slowly than other parts of the UK whereas London, the East of England and Yorkshire and The Humber are growing more rapidly. The English region with the largest positive growth in 2022 was London, at 1.2%, while the largest negative growth was in the North East, at -1.6%.

The richest part of the UK is London. In 2020, the capital had the highest gross disposable household income (GDHI) per head with each person having £29,890 available to spend or save. Northern Ireland had the lowest GDHI at £17,301, well below the UK average of £21,440. The places with the highest GDHI per head in 2020, were the London Boroughs of Kensington and Chelsea on £60,277, and Hammersmith and Fulham. The place with the lowest GDHI per head at £13,952 was Nottingham in the East Midlands.

On average people living in the south live longer than those in the north. At national level, male life expectancy in 2020 was highest in England at 79.4 years and lowest in Scotland where it was 76.8 years. Within England, sizable regional differences are present, including a three-year gap between the North East at 77.6 years and the South East where male life expectancy was 80.6.

For females, country and regional differences were smaller. Between countries life expectancy was highest in England at 83.1 years and lowest in Scotland where it was 81.0. The regions with the highest female life expectancy were London at 84.3 years and the South East at

84.1. The lowest were in the North East and North West at 81.5 and 81.7 years.

In general, children educated in the south achieve better results than those in the north although there is not a clear north-south divide. In 2022, 6.8% of all students in all GCSE subjects in England were awarded a Grade 9. However, there were regional differences. Buckinghamshire scored above the national average with 11.3% of students awarded a Grade 9. Students in Surrey and Cambridgeshire also achieved more Grade 9s than in other counties. In the Isle of Wight only 3.4% of students were awarded a Grade 9, and in the east Riding of Yorkshire and Staffordshire the figure was 3.9%.

There are several strategies that are used in an attempt to resolve regional differences. These include investment in public services such as health care and education, as well as building new transport infrastructure like HS2. Government departments have also been moved out of London. For example, the Drive and Vehicle Licensing Agency (DVLA) is now in Swansea.

The UK government also has a £4.8 billion Levelling Up Fund which provides money for projects which will make a difference to their local economies. For example, in 2023 it was announced that Aberdeen has been awarded £20 million from the fund. The money will be spent on the creation of a new market in the heart of the city centre. There will be better connectivity between the north of the marketplace and local transport, and the streetscape along Union Street will be upgraded. This will encourage more people to shop in the city centre. The new market will have a variety of stalls, some temporary and others permanent, and will have space to celebrate international food and drink festivals.

Another example is Newcastle which has been awarded £7 million from the Levelling Up Fund to secure the future of one of its most important cultural assets, the historic Grainger Market. The Grainger Market is housed in a 200-year-old Grade 1 listed building and is located in the original retail heart of Newcastle. It is one of the few places that sells fresh and affordable produce in the city. The market will be restored and upgraded so that the customers and traders will have a better experience. The improvements will attract new small businesses, encourage greater footfall and create more retailing and job opportunities.

The place that the UK occupies in the wider world is that of a high-income European economy with close links to the rest of the world. The country is a global leader in financial services and foreign affairs and has leading roles within NATO, the Commonwealth and the United Nations.

Trade is the buying and selling of goods and services between countries. Trade is important to the UK, as the country is an important trading nation and is the world's sixth-largest importer and exporter. In 2022 the UK exported goods and services valued at £803 billion and imported £884 billion worth. This resulted in a trade deficit of £81 billion. The UK's largest trading partners were the United States of America, Germany, the Netherlands and China.

It is estimated that 1.35 billion people speak English worldwide which has enabled British culture to have a global influence. In music, literature, theatre, television and the movies, British musicians, writers, and actors have achieved worldwide recognition. The Beatles, Charles Dickens, Harry Potter, Shakespeare, Dr Who, and James Bond have all made an impact in other countries. Britain has also been influential in the world of sport. The English Premier Football League has a global audience.

In addition, the UK benefits from the soft power yielded by both the Royal Family and the BBC. King Charles III is the Head of the Commonwealth. The BBC is the largest broadcasting corporation in the world, respected for the accuracy of its reporting and the quality of its entertainment programmes.

The UK has an important role in global transport networks because Heathrow is a hub airport. In 2019, 82 airlines operated at Heathrow, serving 206 destinations across 86 countries, and carrying 81 million air travellers many of whom were transfer passengers catching a connecting flight to another destination.

The UK also has well developed electronic communications with the rest of the world. Undersea cables connect Britain with other countries. For example, the 6,250 km long Grace Hopper cable runs under the Atlantic. This Google owned cable connects New York in the US to Bude in the UK and Bilbao in Spain. The Amitié cable is another submarine cable which connects Massachusetts USA, with Le Porge in France and Bude in Cornwall. It is owned by a consortium which includes Meta, Microsoft, and Vodafone.

At the national level, the government has pledged to build a world-class digital infrastructure and to roll out full fibre broadband to the whole country by 2033. This will enable the digital economy to expand as people and businesses will have high speed access to the internet through their computers and mobile devices.

The UK has economic links which take the form of trade agreements with a number of different countries. Trade agreements set out the rules for buying and selling goods and services between two or more countries. They reduce restrictions on imports and exports, which makes trading easier and cheaper. As an independent trading nation, the UK now has over seventy trade agreements in place. These include a Digital Economy Agreement with Singapore, which reduces the cost of trade between the two countries by adopting digital trading systems between businesses and consumers. Agreements also exist with Japan, Norway, Iceland, Liechtenstein, Australia and New Zealand.

The UK government is currently negotiating to join the Gulf Cooperation Council, whose members include India, Canada, Mexico and Israel. Agreement was reached in March 2023 for the UK to join the Comprehensive and Progressive Agreement for Trans-Pacific Partnership (CPTPP), a free trade area with Canada, Australia, Brunei, Chile, Japan, Malaysia, Mexico, New Zealand, Peru, Singapore and Vietnam. This will lower trade barriers to the Indo Pacific, an increasingly important global region.

The European Union (EU) is an international organisation and economic and political union between twenty-seven European countries and a community of five hundred million people. Member states have unrestricted access to the European single market, which means that goods, services and people can flow freely across national borders increasing economic activity and trade. The EU was formed to reduce trade barriers and increase cooperation amongst its members. Twenty members of the EU use the euro as their currency. A person who is a citizen of an EU country can live and work in any of the other member states without needing a work permit or visa.

The UK was a member of the EU between 1973 and 2020. Since leaving the EU and becoming an independent sovereign state again, Britain has regained control of its laws and borders and no longer contributes £8 billion a year to the EU budget. The UK government has taken back responsibility for trade, regional development, transport and agriculture. However, Britain has retained strong trading links with the EU and has signed a UK-EU Trade and Cooperation Agreement treaty. In spite of this, the value of the UK's exports of goods and financial services to the EU has fallen. Instead, the UK is trading more with the rest of the world and has an independent seat at the World Trade Organisation.

The UK also has economic and political links with the Commonwealth. The Commonwealth is a voluntary association of 56 independent and equal sovereign states, including Canada, Jamaica, Ghana, Nigeria South Africa, Cyprus, Australia and New Zealand. The smalles member is the Pacific nation of Nauru with a population of 10,000 and the largest is India with 1.4 billion people. Most Commonwealth countries were territories of the former British Empire. The Commonwealth is home to 2.5 billion citizens – one-third of the world's population. Member states have no legal obligation to one another. Instead, they are united by language, history, culture, and thei shared values of democracy, human rights, and the rule of law.

The work that the Commonwealth undertakes varies from protecting the environment, boosting trade, supporting democracy and the rule o law, to developing young people through education and sport. Expert help countries prepare for bilateral and multilateral trade talks so tha they can export more of their goods and services and become mor globally competitive. By working together, member countries facilitat economic growth, peace and prosperity.

The UK plays a leadership role within the Commonwealth b supporting different projects. For example, in 2022 the UK provide £60 million to the Girls in Rwanda Learn (GIRL) programme, and £13 million to GOAL's Action for Learning Project in Pakistan. The UK als backs women-owned businesses through the She Trades programme which provides financial support to over 3,500 female entrepreneurs i Bangladesh, Ghana, Kenya and Nigeria.

In these different ways, the UK is able to exercise its soft power an influence in far flung corners of the world.

17. The challenge of resource management

F ood, water and energy are fundamental to human development. The economic and social development of a country depends on there being reliable supplies of clean water, energy and food. Countries without these resources develop at a slower rate than places with plentiful supplies. Resource management is the control and monitoring of the provision of food, water and energy so that they do not become depleted or exhausted.

The significance of food, water and energy to economic and social well-being is well understood. Well-being is the state of feeling comfortable, happy and healthy. People who live in well-resourced countries with good supplies of nutritious food, clean water and reliable energy have higher levels of well-being than people who live in countries with food, water or energy insecurity.

Countries, which are well-resourced, have a better quality of life with higher incomes, longer lives and higher literacy rates. Reliable supplies of good food, clean water and energy enable people to be healthier and more economically active. This enables a country to become wealthier and more prosperous. Well-resourced countries are also usually better prepared for challenges such as natural hazards, adding to the feeling of well-being.

In the UK, a higher-income country (HIC), people have high levels of well-being because there are good supplies of food, water and energy. In general, UK residents spend a lower percentage of their incomes on food, water and energy than citizens of a lower-income country (LIC). British people also have higher disposable incomes and can afford clothes, electrical items and entertainment, all of which increase their sense of well-being.

In low-income countries (LICs) levels of well-being are lower. For example, in Ethiopia and Eritrea there is a low sense of well-being because of high food prices, shortages of clean water, and unreliable energy supplies. Households spend a higher percentage of their income on these resources which means that they have less to spend on housing, healthcare and transport, which are all things that can add to their sense of well-being. Even in a middle-income country (MIC) like Egypt, well-being can be negatively impacted by poor supplies of clean water.

An overview of the global inequalities in the supply and consumption of resources shows that some countries not only have a lot of resources but also have high levels of consumption.

There are global inequalities in the supply of food. The US (a HIC) and China, Brazil and India, which are all MICs, are the world's top four food-producing countries. All four nations have fertile farming land and climates suitable for the growing of a variety of crops. China and the US produce half the world's supply of maize. China and India produce 50%

of the global supply of rice, although much of India's output is produced by subsistence farmers and is consumed locally. The US is the world's largest exporter of food particularly grain but Asian countries grow more cereals, rice and sugar, and produce more meat, milk and fish than other parts of the world. European countries grow the most barley and are the second biggest producers of some types of meat, milk and fish.

Countries with poor supplies of food include Chad, the Democratic Republic of Congo, Madagascar, Central African Republic and Yemen, all of which are LICs. Food supplies are low in these countries because of civil war and conflict, exacerbated by the effects of climate change, poverty and poor government.

There are also global inequalities in the supply of water. Some countries, for example Indonesia (a MIC) have good supplies of fresh water because they are located along the Equator where the annual rainfall is 4 metres. Other countries with a lot of rain are found in the higher latitudes, and include HICs such as the US and Canada, and Russia and China, both of which are MICs. Places with low rainfall are found along latitudes 30°N and 30°S, and include Sudan, Niger and Somalia, all of which are LICs. In these countries the climate is very dry with less than 250 mm of rain per year. Many other LICs have both low levels of rainfall and poor water infrastructure, resulting in a shortage of clean drinking water and poor sanitation.

In addition, there are global inequalities in the supply of energy. Non-renewable energy comes from fossil fuels which are natural fuel such as coal, petroleum and natural gas, formed in the geological past from the remains of living organisms. These finite resources will eventually run out. Some countries have large amounts of these resources and these include China, the US, Russia and Saudi Arabia. Other countries, for example Japan, have either small amounts of fossil fuels or none at all and rely on imports of oil and gas for their energy supplies. However, the shift towards using renewable energy, will mean that in future most countries will be able to generate the energy they need from renewable sources such as wind, solar, and hydro power.

There are global inequalities in the consumption of food, which is measured in calories, a unit of energy. Generally, as a country becomes richer, its citizens consume more calories. In 2022, the highest number of calories at 3,540 a day were consumed by adults in North American and European countries. In Africa the average daily consumption was 2,600 calories. This is because households with larger disposable incomes can afford to buy not only more food but richer foods containing more calories.

The global consumption of water is continuing to rise. The countries which consume the most include China, the US, Brazil, Russia and India. This is because they have large populations and high rainfall. The countries which consume the least water are mainly African ones, and include Eritrea, Uganda and Ethiopia, all of which are LICs. Not only do these countries have low rainfall and poor water infrastructure, but also low levels of economic development.

There are also global inequalities in the consumption of energy. Some countries consume far more energy than others. Countries in the Asia Pacific region, including China and Japan, consume the largest amount of oil because they have higher levels of development and rising incomes, enabling people to buy more cars, household appliances and electronic devices. In contrast, countries in Africa consume only 4% of the world's oil and small amounts of other sources of energy due to the low incomes and rates of development.

The changing demand and provision of resources in the UK creates both opportunities and challenges. As the population of Britain rises, and standards of living improve, the demand for food, energy and water resources increases. People have higher disposable incomes, which they an spend on a wider variety of food, better housing, and more consumer goods as well as travelling more often, all of which increase their consumption of resources and their carbon footprint. Overall, the UK is well supplied with both home-grown and imported food, fresh water and non-renewable and renewable energy.

An overview of food resources in relation to the UK includes the growing demand for high-value food exports from low-income countries, the all-year demand for seasonal food and organic produce, larger carbon footprints due to the increasing number of food miles travelled, the moves towards local sourcing of food, and the trend towards agribusiness.

The growing demand for high-value food exports from low-income countries happens because Britain is not self-sufficient in food. In total the country produces 54% of the vegetables, 16% of the fruit, and 86% of the beef it consumes, and imports the rest from abroad. For example, onions, tomatoes, and cucumbers come from Spain and the Netherlands, grapes from South Africa and beef from Ireland. Overall, a huge variety of other foodstuffs are imported from over 160 countries located in Europe, Africa, North America, Asia, and Australasia.

The all-year demand for seasonal food is one of the changes in consumer preferences due to rising disposable incomes. Supermarkets are now stocked all year round with high-value foods such as coffee, sugar, wine, fresh and dried fruits and vegetables, and nuts. Many of these foods are imported from lower-income countries. For example, The Gambia exports peanuts, and Madagascar exports rice, sugar cane and mangoes.

As the UK strawberry season runs from May to October, at other times of the year fresh strawberries are flown in from other countries such as Mexico. British apples are only in season from July to September so the supermarkets stock apples from other countries such as New

Zealand during the winter months. More people are also eating organic produce which is grown without the use of man-made fertilizers o pesticides. Crop rotation, animal and plant manures, hand weeding and biological pest control are used instead.

The demand for organic produce has also increased in recent years Organic food is food which is produced using environmentally and animal friendly farming methods on organic farms. Artificial fertilizer are banned and farmers develop fertile soil by rotating crops and using compost, manure and clover. Organically produced food must be free of synthetic additives like pesticides and dyes.

Carbon footprints are a measure of the amount of greenhouse gase produced by a country or a person. A carbon footprint is expressed a tonnes or kilograms of carbon-dioxide equivalent. The UK has a large carbon footprint than before partly due to the increasing number of food miles that imported food has travelled. A food mile is the distance between where an item of food is grown or produced to where it i eaten. When food is transported to the UK, carbon is emitted by the fossil fuels used to power the ships, airplanes and lorries. So imported food has a higher carbon footprint than produce grown locally.

The UK's carbon footprint has also increased because of the demand for high-value foods such as beef and salad crops. Cattle emit methane which is a greenhouse gas, and carbon dioxide is emitted when tomatoe and cucumbers are grown in heated greenhouses.

Local food sourcing is a method of food production and distribution that is undertaken close to where the consumers live. Locally produced food has lower food miles as the distance needed to transport the food to the shops is much shorter. Moves towards local sourcing of food i occurring in an effort to reduce food miles and the environmental impacts of selling imported food. Consumers are also becoming more aware of where their food has come from and are buying directly from local producers either at farm shops or at farmers' markets. The food they buy is fresher, and has lower food miles and a smaller carbon footprint. Farmers also benefit from selling directly to their customers as they can increase their incomes. However, the disadvantage of only sourcing food grown locally is that not all food items will be available all year round, and food prices might be higher due to the higher costs of production in the UK.

Farmers are also seeking ways of reducing the environmental impact of food production. Some have switched from producing beef and lamb to chicken or pork because these animals emit less greenhouse gases. Others are growing more vegetables and cereals.

The trend towards agribusiness is increasing as farms become larger and more commercial. Agribusiness is the application of business skills to agriculture. Farms which are run as agribusinesses use intensive methods with high inputs of fertilizers and pesticides and the use of modern machinery and technology. The advantages of having more agribusinesses are that food security improves, food prices are lowered, farmers make higher profits and there are more well-paid skilled jobs available in agriculture. The disadvantages of agribusinesses are that they force smaller farms with higher costs out of business and the intensive farming methods they use can damage soils and wildlife habitats.

An overview of water resources in relation to the UK includes the changing demand for water, water quality and pollution management, matching supply and demand – areas of deficit and surplus – and the need for transfer to maintain supplies.

The changing demand for water in the UK is complicated by the uneven distribution of rainfall across the country. The wettest areas are concentrated in the west, in the mountainous regions of Snowdonia, the Lake District and the Scottish Highlands, which all receive more than

4,000 mm of rainfall a year. The driest part of the country is eastern England where East Anglia has an average of just 626 mm of rain per year. However, in spite of this uneven distribution all parts of the country are supplied with clean water.

Water is an important resource used daily for drinking, washing, cleaning, and irrigating crops. Since 1985 the population of the UK has grown by 20% but the demand for water has increased by 70%. The average British person uses around 160 litres of water every day, twice the amount that people in Bangladesh use.

This rise in demand for water is because disposable incomes and standards of living are higher than fifty years ago. More people can afford to have their own homes rather than living with other family members or friends. There has been an increase in the number of ensuite bathrooms and a rise in the number of households owning modern water consuming appliances like dishwashers, washing machines and power washers. Standards of hygiene are also higher so people shower and bathe more frequently and wash their clothes more often.

More water is used in both agriculture and industry, because of the increased use of irrigation and the need for water in food processing, electricity generation and manufacturing.

Water quality and pollution management are important issues. Water quality in the UK is of the highest standard, amongst the best in the world. However, on average there is one incident of water pollution every day and the management of pollution is a challenge. It is estimated that half of the UK's rivers are polluted with a cocktail of pollutants, including phosphorous, microplastics, slurry, car tyre particles, oils and wet wipes.

This water pollution affects the natural environment and the food system, as well as impacting on human health and the survival of wildlife. Strong action is taken by the Environment Agency against those who break environmental regulations, but the cost of cleaning up pollution incidents is high and adds to water bills.

Although there has been a clear improvement in serious pollution incidents since 2001, with the UK's rivers the cleanest since records began, the future situation is only likely to get worse with population growth and climate change.

Matching supply and demand for water, while areas of deficit and surplus exist, is a challenge. Currently, some regions have a water deficit which means that demand exceeds supply at times. These areas include London and south-east England, which has low annual rainfall and a large population. In contrast, north-west Scotland, the Lake District, west Wales and south-west England have a water surplus, with more water than is needed, because they have high annual rainfall and lower populations.

During long hot dry spells, hosepipe bans and standpipes are used to limit water usage, but as the number of people grows, ways to increase the supply of water in order to keep up with demand will have to be found.

The need for transfer of water from areas of surplus to areas of deficit has been recognised as a way to maintain supplies. Water is already moved from Wales to the Midlands, and in north-east England, water is transported from the Kielder reservoir in the Pennines, to the rivers that flow south to the cities of Newcastle-upon-Tyne and Middlesbrough. Another solution would be building a national water grid, which would link up the country's water supplies. This would be expensive and technically challenging but could solve the problem.

An overview of energy resources in relation to the UK includes the changing energy mix with less reliance on fossil fuels, the growing significance of renewables, the reduced domestic supplies of coal, gas and oil, and the economic and environmental issues associated with exploitation of energy sources.

The energy mix is the range of energy sources of a region or country, both renewable and non-renewable. In the UK, there is a changing energy mix with less reliance on fossil fuels. In recent decades, Britain has moved away from finite resources such as coal, oil and gas towards more use of renewable energy from infinite sources. In 2022, at least 38% of Britain's power was generated by gas-fired stations, but a record amount of electricity was also produced from renewable sources. This was made up of 27% from wind, 16% from nuclear, 5% from biomass, and 4% from solar. Output from these non-fossil fuel sources was enough to meet national demand for 50 hours.

Renewable energy is unreliable at times – for example when the wind does not blow or temperatures are too high for solar panels to operate. Currently, renewable energy does not meet all domestic demand for electricity due to problems with storage and the fact that not all the different sources of renewable energy are connected to the National Grid. A new generation of batteries should solve some of these problems but it is likely that the UK's energy supplies will be partially met by fossil fuels for the foreseeable future.

In the past, the country benefitted from its large reserves of fossil fuels. In Scotland, England and Wales there were many working coalfields, which powered the industrial revolution in the UK, and led to rapid economic development. The North Sea oil and gas industry began in 1967, with the production of natural gas. The offshore production of oil started in 1975, with the opening of a small number of very large fields including Forties and Brent, and the building of oil and gas pipelines from the oil fields to the shore.

The reduction in the domestic supplies of coal, gas and oil has occurred in recent years. The UK's coalmines were closed because the coal was exhausted, or too uneconomic to mine because the coal is deep underground and therefore difficult and expensive to extract. The use of coal has now fallen to its lowest level since the eighteenth century. At present, the UK's oil and gas comes from over 300 small fields that are more technically complex to produce from, pushing up costs. However, production is increasing and new fields are still being opened. The use of renewable forms of energy has been increasing every year as the UK undergoes an energy transition away from using fossil fuels as sources of power. This is because domestic supplies of oil and gas are running out and burning fossil fuels emits greenhouse gases which contribute to climate change. There is also a need to generate more energy in order to meet the growing demand for electricity.

In Britain, only 2% of electricity is currently generated by coal, which is much lower than the rest of the world. Natural gas is used to produce most of the UK's electricity, but this will reduce as more types of renewable energy become available.

There are economic issues associated with the exploitation of energy resources. Fossil fuels, which are found underground sandwiched between layers of sedimentary rock, are expensive to extract. The more accessible reserves are mined first, but the cost of extraction increases as resources dwindle and deeper seams are tapped. At one time over one million people were employed in the coal industry, but competition from other countries with large open cast coal mines and lower costs, forced the British coal industry into decline, with the loss of many jobs. Drilling for oil and gas can be dangerous with risks of fires and accidents, but the UK's oil and gas industry employs over 200,000 people, generates wealth, and boosts the economies of towns such as Aberdeen. Relying on imports of oil and gas from other countries can be expensive and subject to external shocks such as natural disaster and conflict. Oil and gas power plants also cost money to run and maintain.

Switching to renewable types of energy also has costs. Building onshore and offshore wind farms, solar farms, nuclear power stations, hydro-electric plants and geothermal power stations can all be expensive. A new nuclear power plant can cost £5 billion, although the next generation of reactors promise to be cheaper to build and run, as well as being safer. Building new nuclear plants creates jobs and boosts the local economy, but decommissioning old ones is expensive. One of the advantages of nuclear power plants is that they produce electricity for over sixty years. Another is that they use uranium, which is a relatively low-cost fuel.

Wind power has become one of the largest sources of energy in the UK. In 2020, the UK generated 76,000 gigawatt hours of electricity from both offshore and onshore wind. Offshore wind accounted for 13% and onshore wind for 11%. The largest offshore wind farm in the world is located off the coast of Yorkshire. Although wind farms can be expensive to build, they are likely to become cheaper over time. Today the sector employs over 13,000 people and is boosting the economy of some areas, for example the Humber estuary. Householders can benefit from lower energy bills if they are supplied with electricity from a local wind farm. In addition, it is estimated that the use of wind energy reduces carbon dioxide emissions by over 32 million tonnes every year.

However, wind farms only produce electricity when the wind is blowing, which is sometimes for only 50% of the time. Also, poor electricity grid infrastructure in the UK means that wind power is often wasted especially on very windy days when consumer demand for electricity is low. Wind farms in Scotland can produce 10 gigawatts of electricity on peak days, but the grid has the capacity to transport just 6 gigawatts.

Unlike Iceland, the UK currently has no geothermal power plants, although there is a possibility that some will be built in Cornwall.

Solar power is another type of renewable energy that is being used more often in the UK. Using sunshine to generate electricity is sustainable as there are no carbon emissions. However, solar farms can be expensive to build and the solar panels age and have to be replaced. Electricity can only be produced during day light hours. If solar panels are located on farming land that could be used for producing food, then this could impact on food security.

Hydro-electric power plants are found in areas of high rainfall, steep slopes and impermeable rock such as west Wales. They generate 2% of the UK's electricity and have a lifespan of over a century.

There are also environmental issues associated with the exploitation of energy sources. Coal mining led to a lot of damage to the landscape with the destruction of habitats and the mounds of spoil from the mines. When the mines closed, many were landscaped and used for housing, recreation or nature reserves.

Large coal and gas fired power stations used to dominate the landscape. Now, they are being closed, demolished and the land they stand on reclaimed for other uses. The burning of fossil fuels led to smog, carbon emissions and global warming. Since the reduction in the use of fossil fuels, there has been an improvement in air quality and a fall in emissions.

Although nuclear power has low carbon emissions, environmentalists are concerned about harmful radioactive leaks and the safe storage of radioactive material. The release of warm waste water into local rivers can harm local ecosystems. These concerns are taken seriously and add to the costs of operating a nuclear power plant.

Wind farms do not produce carbon emissions and so help reduce the UK's carbon footprint. However, wind turbines have an impact on the landscape and wild life particularly in areas of outstanding natural beauty. They create a noise and can provoke strong feelings from locals who feel they have a negative impact on landscape quality.

Geothermal power has a low above-ground impact but sometimes releases underground naturally occurring radioactive materials such as radon into the air.

In conclusion, it is impossible to avoid the environmental impact of using energy.

18. The challenge of food management

The demand for food resources is rising globally but the supply of food can be insecure, which may lead to conflict. Although experts do not agree, it has been estimated that the current demand for food could increase by 60% by 2050, and by 80% by the end of the century.

However, the amount of food available depends on the weather. Crop yields fall when the weather is unfavourable, with either too much or too little rainfall or temperatures which are too high or too low. When major food producing countries such as Russia and Australia are hit by droughts, floods, heat waves or cold snaps their exports of wheat and barley fall, leading to world shortages and rising prices. The amount of food produced can also be lowered if crops and livestock are affected by pests and diseases. Shortages of food may lead to conflict when different groups compete for control over the best farming land or access to water for irrigation.

Globally there are areas of food surplus and security as well as food deficit and insecurity. There are also patterns of calorie intake and food supply. The reasons for increasing food consumption include economic development and rising population. The factors affecting food supply include climate, technology, pests and disease, water stress, conflict, and poverty.

Some countries have a food surplus and food security. Food security means that the people have access to sufficient safe and nutritious food, at affordable prices, at all times. This ensures that their dietary needs and food preferences are met, and that they can live a healthy active life. All higher-income countries (HICs) including the UK, the US and Singapore have food security.

Other countries have a food deficit and food insecurity, which means that for most of the time, many people are without reliable access to a sufficient quantity of affordable, nutritious food. Many lower-income countries (LICs) such as Madagascar, the Democratic Republic of Congo, Somalia and Haiti live with food insecurity all the time.

There are uneven global patterns of calorie intake. Calorie intake is higher in the HICs of Europe and North America where more animal products, such as meat, milk and eggs are eaten. Generally, calorie intake is lower in LICs. For example, Sub-Saharan African nations such as Mozambique consume fewer calories than higher income countries. This is because corn, wheat and rice are the staple foods in many LICs. These foods contain fewer calories than other foods such as meat. However, calorie intake is increasing more rapidly in LICs than in other parts of the world as living standards improve and more people can afford to eat meat.

The global pattern of food supply is also uneven. In the HICs food supplies are reliable and plentiful with not only home-grown food but also an array of imported food. In the LICs food supplies can be low and unreliable due to poor harvests, drought, poverty and conflict.

The reasons for the increasing food consumption around the world include economic development and rising populations. When a country becomes economically developed and Gross National Incomes (GNIs) rise, more food is consumed. This is not only because farmers can afford to use mechanisation to increase productivity but also because of improvements in storing and transporting fresh and frozen foods, so that food is available all year round in supermarkets.

When disposable incomes increase people switch to a more westernised diet which includes foods higher in calories such as processed foods, meat and dairy products. However, there are exceptions to this. Although Japan is a HIC, calorie consumption has not increased in line with the rise in GNI because the traditional Japanese diet of healthy whole-foods is low in calories. Some middle-income countries, such as Morocco, have a higher food consumption than expected, because they have embraced western diets. In general, food consumption is lower in LICs because most people eat a staple diet of cereals, which are lower in calories than meat.

The overall consumption of food is increasing worldwide because of rising populations. In 1915, there were 1.8 billion people in the world but today there are 8 billion. It is estimated that by 2050 the global population will have risen to 9.7 billion, with more than half of this growth expected to occur in Africa. The challenge of resource management will be to continue to increase global food production in order to feed everyone.

There are many factors including climate, technology, pests and diseases, water stress and conflict which affect food supply. The climate determines the type of farming that can be practised and the crops that can be grown in a country. For example, rice is widely grown in south-east Asia because of the tropical climate. However, bad weather can reduce crop yields. If there is a poor harvest, then the supply of a particular crop can be lower that year. In 2020, the UK's wheat harvest was less than normal and the worst for 30 years because of a wet stormy winter followed by a dry spring and a hot summer. In early 2023, cold weather and heavy rain in Spain and Morocco affected the ripening of tomatoes, resulting in a shortage on the shelfs of British supermarkets. Extreme weather events such as droughts, floods and heatwaves brought on by climate change can also lower crop yields and affect food supplies.

A lack of technology can result in less food being produced. In LICs, yields are lower and food supplies more unreliable, because farmers rely on manual labour, as well as simple tools and techniques.

Pests are another problem – they can reduce the supply of food by lowering crop yields and farm outputs. Not only can pests cause huge devastation but they can be difficult to control. In Africa, swarms of desert locusts eat a very wide range of crops. They can fly in unexpectedly and strip a field bare in an hour. In Mozambique, rodents such as rats and mice eat not only the crops growing in the fields but also food kept in storage. In addition, they cause food poisoning by contaminating food and water supplies with their urine and droppings. Rats can spread fever and other diseases by biting people and animals.

Diseases can also affect food and drink supplies. In the 1980s, a fungal disease called witches' broom infected Brazil's cocoa trees. Many farmers saw their harvest of cocoa beans fall by over half. Brazil lost its position as one of the top global producers of cocoa. In summer 2023, mildew affected the merlot grape vines in many of the vineyards of the Bordeaux region of France. Some farmers lost almost all their grapes to the disease. As a result, there were many fewer bottles of red wine produced that year.

Water stress occurs when there is insufficient water available to meet the needs of farmers and the general population. Many countries with the highest risk of water stress are in the Middle East, such as Kuwait and Qatar. Singapore is also one of the most water-stressed countries in the world due to a lack of natural water resources. The country has long relied on water supplied by its neighbour Malaysia.

Conflict can also affect food supplies. If fighting is taking place in an agricultural area, crops may be destroyed by invading armies. Farmers may be prevented from sowing or harvesting their crops or may be forced to flee to safety. Some farmers may even leave their farms to join the fighting. This happened in the Central African Republic, a LIC, in 2017 when armed conflict forced over one million people to leave their farms and homes.

Hostile groups can deliberately cut off food supplies to a region, in order to weaken the inhabitants. The war in Ukraine, which started in 2022, affected the global food supplies of wheat and sunflower oil, causing shortages of food, particularly in the Middle East and Africa.

Poverty also reduces food supplies. In many LICs, such as the Central African Republic where 80% of the 5 million people live in poverty, food production is low because farmers cannot afford to buy the pesticides or farm machinery they need. Also, when there are food shortages, the government lacks the funds needed to import food, particularly when global food prices are high.

The impacts of food insecurity include famine, undernutrition, soil erosion, rising prices, and social unrest. Famine occurs when a large number of people are affected by a widespread shortage of all types of food. The worst famines can lead to starvation and even death. Recent famines have occurred in 2016 in Yemen, in 2017 in South Sudan, and in 2021 in Madagascar. The effects of famine include undernutrition and death through starvation.

Undernutrition occurs when people eat a diet that lacks variety and has a low nutritional value with a shortage of calories, vitamins and minerals. Undernourished people do not have enough nutrients to cover their needs for energy and growth, or to maintain a healthy immune system. Their resulting poor health affects their ability to work and earn money. Children who are undernourished become underweight and suffer from stunting and wasting. Undernutrition is found in LICs such as Burundi, Kenya and Madagascar.

Food insecurity can sometimes lead to soil erosion during a period of population growth. To feed more mouths, marginal land is cleared of woodland and scrub, and brought into cultivation. Existing pastures can be overgrazed and fields overcultivated. These poor farming practices can result in the soil becoming infertile, dry, and more susceptible to erosion. This has occurred in northern Sudan, where overgrazing and the expansion of the cultivated area, has led to widespread soil erosion and desertification.

Rising food prices in markets and shops can also occur as a result of food insecurity. In 2020, the price of grain, meat, milk and bread rose in Sudan, because of shortages in these basic foods. This hit the poorest hardest. However, some farmers and shopkeepers benefited from being able to charge higher prices. If food prices stay high for a long period, the standard of living of the people will start to decline, as more money is spent on food and less on housing, healthcare and education.

Social unrest is another impact of food insecurity. When people are unhappy about rising food prices they gather to protest. These protests against the government can become angry and violent. In 2003, in the Darfur region of western Sudan, conflict broke out over land and grazing rights. In 2020, street demonstrations continued for months after the Sudanese authorities tripled the price of bread.

So how can food supply be increased? There are several different strategies that can be used and these include: irrigation, aeroponics and hydroponics, the new green revolution, the use of biotechnology, and the use of appropriate technology.

Irrigation is applying water to land in order to supply crops and other plants with necessary water in order to help them grow. Farmers can use water from a variety of different sources including from a spring, well, lake or reservoir. Or they can use desalinated or recycled grey water. There are several different methods of irrigation. Surface or gravity irrigation is one of the oldest methods and is still widely used. In this method, water runs down from the top of a field to the bottom of the slope in a series of shallow channels. In drip irrigation, a small amount of water drips from a water pipe onto the soil around a plant. With the help of irrigation, crop yields can be doubled. In many parts of Africa irrigation is still not used so this would be an easy way to help produce more food.

Aeroponics involves growing plants without the use of soil – just with air, water and nutrients. In an aeroponic system, the roots of the plants grow in air and are sprayed with a nutrient solution containing nitrogen, phosphorus, and potassium. Hydroponics is another method of growing plants without soil. In hydroponics, the crops are grown in a water solution containing the same nutrients. Oxygen is pumped into the water to prevent the plants from drowning. Both aeroponics and hydroponics are closed systems.

The advantages of both aeroponics and hydroponics are that there are lower inputs of soil, water and fertilizer, resulting in lower costs. As the plants are not grown in soil, these techniques are particularly suitable for use in places with poor soils. Also, both methods save space and can be used in cities. Yields are higher than in conventional farming.

The disadvantages of these systems are that they are small scale, and require a lot of expertise, technology and maintenance. Seed potatoes are grown using aeroponics by some farmers in Vietnam, and hydroponics is used in some places to grow vegetables in India. In the UK, herbs and salad crops are being grown using these techniques.

Another way of increasing food supply is to use higher yielding varieties (HYVs) of seeds. Farmers first started sowing these seeds in the first green revolution in the 1960s. Today, a new generation of HYVs has led to the new green revolution, which is taking place in many parts of the world. The new green revolution is a combination of modern technology, traditional knowledge and working with nature. Farmers sow the new HYVs but also use new farming techniques and information systems to monitor their crops. In India, the use of HYVs, which are drought, pest and disease resistant, has increased the production of rice by producing larger healthier crops. Environmental damage from external inputs has also been minimised. The new green revolution is particularly benefiting farmers farming marginal areas in LICs.

Biotechnology is the manipulation, through genetic engineering, of living organisms to produce useful commercial products such as pest resistant crops and new bacterial strains. Biotechnology, which is the use of biology and technology, can increase food supply. In biotechnology new varieties of plants are developed using gene editing and genetic modification (GM). The new varieties of GM maize, oilseed rape and soya bean are higher yielding and more reliable, raising farmers' incomes and reducing the use of pesticides and insecticides. Biotechnology is also used in food processing. With the use of sweeteners and flavours in-edible food is manufactured into safe tasty food with an extended shelf-life.

Appropriate technology, which includes simple tools and equipment suited to the needs and resources of the local farmers, can also raise food supply. This is happening in the Sahel region of Burkina Faso, Niger and Nigeria where farmers are using the Delfino plough, which is a heavy digger, to cut through compacted dry soil to a depth of more than half a metre. This enables farmers to extend the area of cultivation and to grow native species of crops, which are better suited to the degraded land than foreign varieties.

Other simple techniques that the farmers of the Sahel use are the Zaï method and the planting of maize under trees. In the Zaï technique, circular holes are dug in the ground which are then filled with water and compost and planted with millet and sorghum. In some parts of the Sahel, maize is grown under leafless Faidherbia trees. These trees do not compete with the crop for light, nutrients or water while the maize is young, but provide shade and protection. These types of appropriate technology increase crop yields and help to improve soil fertility, but are small scale and are not very effective during long droughts.

In the UK, farmers also use appropriate technology. For example, tarpaulin is used to protect crops such as hay and straw from the weather. Trees are often planted as wind breaks, in order to provide some protection from frost and wind.

Another way of increasing food supply is to have larger farms which have lower costs and higher outputs. A large-scale agricultural development is considered to be over 1,000 acres or 400 hectares.

The Elveden Estate is an example of a large-scale agricultural development, which can show both the advantages and disadvantages of large-scale farming. Elveden, in Norfolk eastern England, covers 10,000 acres and is the largest arable farm in lowland Britain. Using powerful machinery on the large 100-acre fields, they grow vegetables, including potatoes, onions, carrots and parsnips, and cereals such as rye, barley and wheat. The farm is managed by a highly skilled team of workers using computer technology, irrigation and modern agricultural machines.

The advantages of farming a farm this size is that the economies of scale have led to the production of a larger quantity of high-quality crops, higher profits for the farm's owner, but also lower food prices in the shops.

The disadvantages of such a large-scale farm as the Elveden estate are that other smaller farms may be forced out of business because they cannot compete on price and food quality. In addition, the intensive farming methods and the large fields may also be less beneficial to the environment and may lead to a loss of biodiversity.

Moving towards a sustainable resource future is occurring in the UK. The potential for sustainable food supplies can come from organic farming, permaculture, urban farming initiatives, fish and meat from sustainable sources, seasonal food consumption, and reduced waste and losses.

Sustainability is concerned with providing enough safe and nutritious food to meet the health needs of the present generation, without compromising the ability of future generations to meet their own needs. A sustainable food supply is good quality safe food that is produced without damaging the soil or water supplies and which makes contribution to the local economy.

Organic farming grows food using only natural fertilizers such as compost, manure and bone meal. To keep the soils fertile, crop rotation and companion planting are used. Organic farming is more sustainable than farming which uses pesticides because fewer artificial chemicals are used on the land and there are lower carbon emissions. However, organic farming produces lower crop yields and has higher costs, making organic produce more expensive to buy.

Permaculture or permanent agriculture is a system of farming that mimics the patterns and features of natural ecosystems. Permaculture is a sustainable system because it is a type of farming that cares for the soil, causes no pollution, and produces a wide variety of different foods.

Urban farming is the growing of fruits, herbs, and vegetables and raising animals in towns and cities, a process that is accompanied by many other activities such as processing and distributing food, collecting and reusing food waste. Urban farmers grow relatively small amounts of food in a variety of locations including on rooftops, balconies and allotments. Urban farming is sustainable because it produces fresh food in season, attracts wildlife and brings communities together. However, urban farming is too small-scale to be able to feed large numbers of people.

More people are eating fish from sustainable sources. In the past most fisheries were overfished, but today many countries including the UK are adopting a more sustainable approach to fishing. The size of the gaps in fishing nets has been increased to ensure that juvenile fish can reach reproductive maturity and have offspring before being caught. Also fishing quotas are used to reduce overfishing. These strategies have led to Dover sole being a sustainable fish to eat, as there are healthy populations along the south-west coast of the UK, in the English Channel and in the Irish Sea.

Meat from sustainable sources is also being eaten more than before. Beef which is sustainable has been reared according to animal welfare standards and in an environmentally friendly way. Cattle are kept in fields rather than indoors and fed on grass rather than grain. No hormones are used and the amount of anti-biotics is restricted. Many people are also eating more turkey and chicken, which do not produce methane and need less food and water than sheep and cows.

Seasonal food consumption is a sustainable practice. This is because food that is in season is fresher, tastier and more nutritious than out-of-season foods. In addition, seasonal foods have smaller carbon footprints because they have not travelled so far. However, relying just on seasonal produce would mean a limited choice of food in supermarkets and more careful menu planning.

Sustainable food supplies can also be achieved by reducing food waste and food loss. It is estimated that one third of all food is wasted. In high-income countries food is thrown away because of cosmetic reasons and short use-by dates. In low-income countries a lot of food is lost due to poor storage. There are various ways of cutting food waste and losses. These include storing food better using refrigeration and climate-controlled warehouses and extending use-by dates.

Rice-fish farming in Bangladesh is an example of a local scheme in an LIC or NEE which is increasing the sustainable supplies of food. In the Jamalpur district of Bangladesh the rice farmers introduce fish to their paddy fields.

To start this type of farming, the farmers build a dyke or bund, which is a raised bank about 60cm high, all the way around one of their rice fields. The dyke not only keeps the fish and water in the field but can also be used as a place to grow vegetables. Next, a ditch is built across the field and filled half full with water purified with lime and enriched with fertilizer. The field is then planted with rice seedlings in rows 35cm apart and flooded with water 12cm deep. Small fish are added to the ditch and when they are used to the environment they are released into the field. As they and the rice grow the farmer raises the water level in the field.

The fish thrive as the rice plants protect them from birds. The fish droppings fertilise the soil which helps the rice to grow. The fish also eat insect pests. Four months later, the farmer harvests the rice first, then drains the field and catches the fish.

There are many positive impacts of this rice-fish culture. There is a 10% increase in rice yields so the farmers have more food to eat as well as having more rice and vegetables to sell at the local market. The increase in the farmers' incomes can be used to buy goods and services which improve the standard of living of themselves and their families.

The fish is eaten by the local people adding a regular source of protein to their diets which improves their health.

Rice-fish culture is a sustainable way of increasing food supplies, because not only is it environmentally friendly, but it uses the skills and expertise of the local farmers, as well as increasing their incomes and improving their diets and health.

19. The challenge of water management

The demand for water resources is rising globally at the rate of 1% a year, which is in line with world population growth. Rising populations and economic development both increase the demand for water. However, the supply of water cannot always be guaranteed. Water shortages sometimes trigger outbreaks of conflict.

Areas of water surplus are places where the supply of clean water is greater than the demand for water. These places have water security, as there is enough clean water available to enable everyone to live a normal healthy life. Everyone has access to a reliable supply of good quality water, which is safe to drink, at all times. This clean water is colourless, odourless and clear. Sometimes however, even in places with a water surplus, periods of water stress occur. For example, during a drought or after a pollution incident, there may be restrictions on the amount of water that can be used.

Areas of water deficit are places where the water supply is not large enough to meet the demand for water. These places have water insecurity, which means that the people living there do not have enough water to maintain their health and productivity.

The global patterns of water surplus and deficit are uneven. Many of the higher-income countries (HICs) of Europe, Asia and North America, including the US, Germany, the UK and Japan, have water surpluses and water security. Many middle-income countries (MICs) along the Equator including Indonesia, the Philippines and Vietnam also have a water surplus. However, places with low annual rainfall such as hot deserts are more likely to have a water deficit and water insecurity. These include many lower-income countries (LICs) in west Africa, like Burkina Faso, Niger and Chad.

The reasons for increasing water consumption include economic development and rising populations. As countries become more economically developed, they consume more water. Water consumption is higher in high-income countries than in the low-income countries. This is because the governments of richer countries have more money to spend on water infrastructure such as pipelines, sewage treatment plants and reservoirs, and can afford to supply water to remote rural areas. As personal incomes increase and living standards improve, water consumption increases. More people can afford to buy domestic appliances, which use water, such as dishwashers and washing machines and also to take longer more frequent showers or baths as their personal hygiene improves.

Population growth also leads to increased water consumption. Today the global population is 8 billion. It is estimated that by 2050 there will be 10 billion people all needing access to clean water for cooking, cleaning and personal hygiene. The challenge of water resource management will be to continue to increase global supplies in order to meet the needs of everyone.

The factors affecting water availability include climate, geology, pollution of supply, over-abstraction, limited infrastructure, and poverty. Climate is the most important reason why some countries have plenty of water and others do not. Regions with high rainfall usually have a surplus of water, whereas those with drier climates have less. Even a small reduction in annual rainfall in arid places reduces the availability of water. During heatwaves, the rate of evaporation increases impacting on water supplies. One of the effects of climate change is that the climate is becoming less predictable. This means that all countries are having to plan for more periods of drought, as well as excess rainfall.

Geology is another key factor which affects water availability. Sandstone and limestone and other permeable rocks can allow water to flow through them. If a permeable rock is sandwiched between two layers of impermeable rocks, an aquifer can form. An aquifer is an underground layer of rock which stores drinking water. This groundwater can be pumped to the surface or accessed through a well or borehole. Some places in the US, for example Nebraska, rely on groundwater for 50% or more of their water supply. However, if the geology of an area is made up of impermeable rocks, such as granite, rainfall cannot soak into the ground, and instead runs off into streams and rivers. In London 80% of the water supply comes from rivers, while the remaining 20% comes from aquifers.

Water pollution can have a significant impact on water availability. Chemicals, excessive nutrients, sewage, waste from agriculture, industry and households, can all pollute rivers, reservoirs and groundwater supplies. In Africa, some LICs, for example The Gambia and Sierra Leone, have a major problem with water pollution.

In Asia, groundwater supplies in Punjab have been contaminated with arsenic, which cannot be removed by boiling or filtration, only by expensive processes like osmosis.

Over-abstraction occurs when water is being used more quickly than it is being replaced. If groundwater is pumped out of the ground faster than it can be replenished over-abstraction happens. This affects water availability, particularly during droughts. When too much water is removed from the ground, water tables sink, wells run dry, and there are higher pumping costs and problems with salt-water intrusion. Lower water tables mean that springs and river beds dry up. In Italy, over-abstraction from the River Po and from the Milan aquifer has led to groundwater levels dropping up to 40 metres in the last 80 years.

Limited infrastructure is another reason why water may not be readily available. Water infrastructure in many LICs is very limited. The lack of water pipes means that water does not reach homes and so, many families rely on water deliveries, which may be unreliable and expensive. There may also be a shortage of reservoirs, aqueducts, and sewage treatment facilities which means that in the rapidly growing cities the supply of clean water has to be rationed. In many rural areas people also have poor access to clean water. For example, in Angola, the supply of clean water is often miles away, causing women and girls to spend many hours a day fetching and carrying heavy jerry-cans back home.

Poverty can also determine water availability. Low-income countries may not be able to afford to buy water from their neighbours in times of shortage. Poorer households may not have enough money to buy bottled water from the shops. Poor countries may lack water storage facilities or have the money to maintain existing ones properly. If wells and storage tanks become contaminated with sewage there is a higher risk of waterborne diseases, which are caused by microorganisms in dirty water. Infection commonly results during bathing, washing, drinking, food preparation, or by eating infected food. Cholera, typhoid, and botulism are all examples of waterborne diseases.

The impacts of water insecurity are both economic and social. The impacts include an increase in waterborne disease and water pollution, a fall in food production and industrial output, and the potential for conflict where demand exceeds supply.

There are clear links between waterborne diseases and water pollution. When there are water shortages, people may have no choice but to drink contaminated water. By doing so, people run the risk of acquiring waterborne diseases, which can leave them unfit to work, thus reducing their incomes. In severe cases, death of the family breadwinner may occur leaving the family destitute.

In many LICs, the lack of a reliable supply of clean water impacts on the standards of living and quality of life of the people, particularly the poor. When water is very scarce, people become distressed, thirsty and dehydrated. This puts an extra strain on health services. Water insecurity can lead to lower standards of personal sanitation and hygiene causing health problems, especially in women and girls. In Eritrea in East Africa, over three quarters of households' drink water contaminated by human waste, resulting in higher levels of illness. In Pakistan, 60% of infant deaths are caused by diarrhoea, as a result of drinking dirty water.

Food production is also affected by water insecurity. Unreliable or low supplies of water mean that there is less available for irrigation. This results in lower crop yields and higher food prices. During a famine there can be widespread malnutrition, illness and death. Many people may have no choice but to migrate in search of food and water.

Industrial output can also be lowered by water insecurity. This happens when there is not enough water to manufacture products. This may lead to factories closing, lower wages, job losses and slower economic development. Labour shortages can also occur if workers are sick with waterborne diseases.

Water conflicts are disputes between different regions or countries about the distribution and use of freshwater. Conflict and social unrest may occur in places where demand for water exceeds supply, and where people share a water source which is diminishing. This can happen if a country extracts more than their fair share of water from a river, or if they propose to build a dam which restricts water supplies downstream. For example, Ethiopia is building the Grand Ethiopian Renaissance Dam and hydro-electric power station on the River Nile. This river development scheme is causing tension with Egypt, as the Egyptians are concerned that their water supplies may be at risk.

Different strategies can be used to increase water supply. The strategies that can be deployed include diverting supplies and increasing storage, building dams and reservoirs, transferring water, and desalination.

Diverting supplies of water involves allowing water to flow from a place where there is a surplus to somewhere where there is a shortage. The water can be used immediately or stored for future use. This can be done on a small or a large scale. For example, gardeners and allotment holders, can fit gutters to their sheds or outdoor buildings, to funnel rain into a water butt or another form of storage. Water companies can divert heavy rainfall and surface water into channels and then store it in underground aquifers. This happens in Oklahoma where there are high rates of evaporation as well as periods of drought.

Increasing the storage of water is another solution. Water can be stored in a variety of different places including in reservoirs, ponds, tanks and water barrels. Both potable water, which is water intended to be drunk, and non-potable water which intended for use in agriculture, can be stored.

Building dams and reservoirs is another solution. A dam can be built across a river in order to stop all the water from flowing downstream into the sea. Behind the dam, the water builds up and forms a reservoir. Dams and reservoirs can also prevent flooding, as the flow of the river is controlled. The water stored in the reservoir behind the dam can be released gradually throughout the year for drinking, irrigation and the production of hydro-electric power.

Dams and reservoirs range in size. Some earth dams are only a few metres high. Others are as large as the Akosombo Dam in Ghana. This dam which is also known as the Volta Dam is 124 metres high and 660 metres long. The Lake Volta reservoir behind the dam, covers 8,502 km². In the UK, the Kielder Dam in Northumberland is 1.2 km long and 52 metres high. Kielder Water, the largest man-made reservoir in Europe, covers 10.86 km² and holds 200 billion litres of water.

However, building and maintaining large river development schemes with a series of dams and reservoirs, can be expensive, destroy habitats, lose farmland and displace people. In addition, there can be problems including the loss of water in the reservoir through evaporation and leaks, and the build-up of sediment behind the dams. However, in many hot arid regions the construction of dams and reservoirs has reduced water insecurity as well as producing much needed electricity.

Water transfer schemes attempt to make up for water shortages by constructing elaborate systems of canals, pipes, and dredging over long distances to transport water from one river basin to another. These schemes are used when a country has a water surplus in one area and a water deficit in another, water supplies can be transferred from one to the other by tunnel, aqueduct, canal or pipe. Water transfer schemes can be small or large scale. An example of a small-scale scheme is the transfer of water from Kielder Water, the UK's largest reservoir, south to the rivers Wear and Tees and then on to the cities of Newcastle and Middlesbrough.

Desalination involves the removal of salt from seawater to produce potable water. Reverse osmosis technology is used to separate the salt from water pumped from the ocean. In the process, salt is trapped when smaller water molecules pass through fine membranes. Although a relatively new way of increasing water supplies, the technology has improved so much in recent years that desalination is now considered to be a reliable source of drinkable water. Wealthy countries such as Saudi Arabia and Dubai, which have desert climates and plentiful supplies of electricity, rely on desalination for their water supply.

Many large cities, including London, also have desalination plants which can be switched on during periods of dry weather. The plant at Beckton takes brackish water from the River Thames and converts it into potable water through a reverse osmosis process. However, desalination is expensive and is an unaffordable solution for many LICs.

An example of a large-scale water transfer scheme, the Lesotho Highland Water Project, occurs between Lesotho and South Africa. Lesotho is a small mountainous country with high rainfall and a water surplus. South Africa, a large country surrounding Lesotho, has a water deficit. The water transfer scheme is the largest in Africa.

The transfer scheme has involved building a network of dams and tunnels between the Lesotho highlands and South Africa. There are several dams including the 185 metres high Katse Dam, which was finished in 1998, and the Mohale Dam which was completed in 2003. Water is collected in the reservoirs behind the dams and then diverted through the 200 km of tunnels to the rivers Senqu, Orange and Vaal.

The water transfer scheme has had several positive impacts. There were political benefits because Lesotho and South Africa worked together on the joint project bringing the two countries closer. Lesotho has benefitted economically from the scheme because the country now receives a regular income from South Africa for supplying the water.

The building of the dams led to new roads being constructed, openin up remote areas. South Africa has benefitted from greater water securit and from the regular supply of electricity. This economic activity led t the multiplier effect with the creation of new jobs in construction an in the water industry.

In South Africa the social benefits from having a more reliable suppl of clean water have been higher standards of living and improved healtl Children now spend less time collecting water and more time in schoo

The water transfer scheme has brought improvements to th environment as well. In South Africa, the water in the Vaal Rive reservoir has been cleaned up by the freshwater from Lesotho. Pollutio caused by the mining industry has also been reduced. Flooding is no\ more under control and the reservoirs are being used for fishing an recreation.

However, the scheme has had several negative impacts. The economi disadvantages include the high cost of the project – over $8 billion higher than originally estimated because of fraud and corruption; an the increase in water bills.

The social disadvantages of this large-scale water transfer schem include the displacement of over 20,000 people, who lost their home when the Katse Dam was built, and the loss of farm land and communa grazing lands.

There have also been some negative impacts on the environment Wetland habitats have dried up because of the lower river leve affecting wildlife and fish. Some species including the Maloti minnow rock catfish and the bearded vulture have become even mor endangered. There have also been technological problems includin leaks and the build-up of sediment behind the dams. However, in spit of these disadvantages overall the scheme has been judged to have bee a success.

Moving towards a sustainable resource future is becoming a reality. This means meeting the water needs of the present generation without compromising the ability of future generations to meet their own. A sustainable water supply is one which supplies safe, reliable, and affordable water, but at the same time minimises adverse effects on the environment and makes sure that future generations will be able to meet their own requirements for water.

There are various strategies that can be deployed to make water supplies more sustainable. These include water conservation, groundwater management, recycling, and the use of grey water.

Water conservation is the preservation, control and development of water resources, both surface and groundwater, and prevention of pollution. In high-income countries like the UK, where a lot of water is wasted, water conservation aims to save water. There are many campaigns to get the ordinary consumer to use less water. This can be done by fixing dripping taps, switching to water-saving washing and cleaning appliances, and spending less time in the shower.

Homes and businesses have been fitted with water meters so that people are aware of the amount of water they are using. Water bills have been raised so that the water companies have more money to spend on repairing leaks and increasing water storage.

In Pakistan various strategies are being used to promote water conservation. The government runs campaigns such as the 2020 World Water Day. The slogan "Wash your hands but Close That Tap" was used to raise awareness of the need to conserve water. Water companies stop illegal connections and repair leaks in order to save water. Farmers are saving irrigation water by watering the roots of crops, by using smart soil moisture sensors, and by building nuccas, which are water channels made with pre-cast concrete.

Groundwater management is the regulation and control of water levels, pollution, ownership and use of groundwater. Groundwater management involves making sure that the water is not extracted faster than it is replenished. The water stored in aquifers is kept as clean as possible with managers ensuring that the water is not polluted by chemicals from mines, fields and factories. They also prevent salt water seeping in from the sea, as decontamination is expensive and can be ineffective.

In high-income countries there is money available to manage groundwater supplies well. In the UK the Thames basin aquifers supply up to 30% of the water used every day across London and the Thames Valley, and help keep the rivers flowing, which supply the remaining water used.

However, in Pakistan the government does not manage the Indus basin aquifer well. Too much groundwater is being used. This over-abstraction means that wells often run dry and have to be deepened. Water supply is unequal with richer families being able to afford their own wells while poorer families have water shortages.

Recycling is another way of saving water. Wastewater and grey water are treated by using machinery and chemicals and then used again in agriculture and industry, but not for drinking. For example, in electricity power stations, steel works and nuclear plants recycled water is used for cooling and industrial processes. On farms, recycled water can be used for irrigation. Using recycled water is more sustainable than using clean mains water because less energy has been used to process it.

Grey water is untreated waste water from domestic baths, showers and washing machines from people's homes that can be recycled and put to good use such as laundry and toilet flushing. Water collected in garden water butts can be used to water lawns and plants but not for drinking and cooking. Treated grey water can be used to irrigate both food and non-food producing plants. The nutrients in the grey water, such as phosphorus and nitrogen, provide an excellent food source for these plants. This is a sustainable use of water because it uses less energy to produce.

An example of a local scheme which is increasing the sustainable supplies of water can be found in Bhatha Dhua, a village in the Ludhiana district region of the Punjab in Pakistan, a lower-middle income country.

To access more clean water, a new tube well which reaches down into the groundwater and can pump up enough water to supply 200 families, has been installed. The water is sold to customers at $3 a month and the money raised is used to cover maintenance costs, electricity bills and staff salaries. This is sustainable because over-abstraction will not occur.

To store more water for use during the dry season, taankas which are large round underground water cisterns fed by rainwater collected from roofs, are being built. This store of clean drinking water can supply a small family for up to six months and can supplement their water supplies during droughts. This provides a sustainable water supply provided the taankas are repaired and kept clean.

The farmers of Bhatha Dhua are also trying to conserve water. They are growing crops on long narrow ridges separated by sloping irrigation channels called furrows. Water flows slowly down the furrows under gravity. Generally, the top of the field receives more water than the bottom, but less water is wasted. They are also building small earth dams, called joheds, along streams in order to store water in small ponds.

Some farmers are using appropriate technology such as drip irrigation, which is a type of micro-irrigation system that allows water to drip slowly to the roots of plants, either on the soil surface or within the soil. This minimizes evaporation and can save three million litres of water every acre.

Others are sowing dry-seeded rice earlier in the year into dry fields, instead of planting young rice plants into flooded fields. The rice matures and can be harvested ten days earlier than usual, creating jobs during a time when there is often a shortage of work. Afterwards the fields can be replanted with potatoes or wheat, increasing food supplies. Although sometimes it is hard to buy the right seeds needed to ensure higher yields, this method saves water, is more productive and creates jobs.

Overall, the strategies being used by the local people of Bhatha Dhua are increasing the sustainable supplies of water all year round, an improvement which is enabling families not only to raise their standards of living but to stay in their villages.

20. The challenge of energy management

The demand for energy resources is rising globally at the rate of 5% a year, which is above the rate of world population growth. Rising populations and economic development both increase the demand for energy. The supply of energy cannot always be guaranteed, and fuel shortages and price rises can trigger disagreements between users and providers. Conflict can also break out if there are unresolved disputes about access to an energy resource such as an oil field.

Energy is needed for a variety of purposes such as fuel for cooking, heating, cooling and running electrical appliances. In addition, most types of transport are still powered by petrol and diesel although new electric vehicles are being introduced. Aircraft use aviation oil and most ships are powered by diesel. Computers and the internet also depend on supplies of electricity. In short, most of modern life relies on energy to function.

The Sun is an important source of energy – providing heat and light during the day. In the past, people burnt wood and dung for heat. Wind, water and animal power were used as well. During the Industrial Revolution, the invention of the modern steam engine meant that coal started to be used on a large scale to power machines and factories. Power lines brought cheap electricity both to the cities and the countryside. In the 19th century, oil started to be processed into petroleum to power internal combustion engines. As cars became cheaper to make and run, urban sprawl with new suburbs became possible. The discovery of North Sea gas in the 20th century led to a network of natural gas pipelines across the UK to heat both homes and offices.

Coal, oil and natural gas are fossil fuels, which were formed in the geological past from the remains of living organisms. Since the burning of fossil fuels releases large quantities of greenhouse gases into the atmosphere, the use of these sources of energy is being phased out and is being replaced with renewable energy sources instead.

Areas of energy surplus are places where the supply of energy is greater than the demand for energy. Some countries, for example Russia and Canada, have an energy surplus. These places have energy security where there are uninterrupted supplies of affordable energy to power agriculture, industry, transport and household appliances. Everyone has access to a reliable supply of energy at all times.

Areas of energy deficit are places where the energy supply is not large enough to meet the demand for energy. For example, Sierra Leone, has an energy deficit with less power available than is needed, and demand is greater than supply. This country and other similar countries have energy insecurity, which means that the people living there do not have enough energy to maintain their normal lifestyles.

In many parts of Africa economic development is being held back by energy insecurity. In many countries such as Kenya and Uganda, there are frequent power cuts which means that businesses cannot rely on the national grid for their electricity supply. Many use diesel generators as a back-up. However, these diesel generators have a high carbon footprint and add to air pollution.

The global distribution of energy supply is uneven. Russia, China and the US have plentiful supplies. However, many sub-Saharan countries such as Malawi and South Sudan – which was ranked as the least-electrified country in the world in 2020 – have low energy supplies. High-income countries which are not energy producers, for example Japan, can afford to import oil and gas and develop nuclear power and other renewable sources.

The global pattern of energy consumption is also uneven. Energy consumption is higher in higher-income countries (HICs) such as the US, and in the newly emerging economies (NEEs) such as China and India. Whereas the majority of lower-income countries (LICs) including The Gambia, Sierra Leone and the Central African Republic consume the least amount of energy.

The reasons for increasing energy consumption include economic development, rising populations, and technology.

Economic development leads to increased energy consumption. As countries become richer, their governments have more money to spend on energy infrastructure including oil and natural gas pipelines, power transmission lines, and energy storage facilities which can supply even remote rural areas. These measures all increase the use of energy.

As consumers become wealthier, they can afford to buy energy-consuming domestic appliances. In HICs, many households use washing machines, dishwashers and tumble dryers, all of which use a lot of energy. In addition, high income families have central heating and air conditioning, and own more than one car.

The rising population of the world is causing an increase in the consumption of energy. In 1915, there were 1.8 billion people in the world but today there are 8 billion. It is estimated that by 2050 the global population will have risen to 9.7 billion, with more than half of this growth expected to occur in Africa. The challenge of resource management will be to continue to increase the production of energy in order to meet the needs of everyone.

Technology is also increasing energy consumption because of industrialisation, mechanisation and modernisation. Digital technology is particularly energy-intensive. The data storage and processing industry already accounts for 4% of global electricity consumption. This is set to rise further as every year the number of smart phone and internet users increases.

The supply of energy is affected by a number of different circumstances including physical factors, the cost of exploitation and production, technology, and political factors.

The physical factors, which determine the amount of energy resources that a country has, include geology, climate, and relief. If there are sedimentary rocks, then it is likely that there will also be some coal, oil and gas. Fossil fuels are found in places which were once shallow seas millions of years ago. When the plants and animals living in the seas died, the dead material sank to the bottom and was covered with layers of sediment such as sand, shale or mud. Over time, under heat and pressure, the sediments were turned into sedimentary rocks, and the organic matter into energy resources. Fossil fuels are found worldwide, but most of the oil and natural gas reserves are located in Saudi Arabia, Russia, the US, and Iran.

The climate is an important physical factor. Solar power depends on the amount of sunshine and daylight – so countries with hot climates can generate larger amounts. Wind power needs large amounts of wind.

Hydro power depends on a high rainfall. The type of relief also influences where hydroelectric power plants can be built, as a highland area with a steep sided valley and impermeable rocks is needed as a suitable site.

The cost of exploitation and of production is a major factor determining whether or not a type of energy is used. Energy exploitation is developing and using energy resources to the greatest possible advantage, usually for profit.

In some places, fossil fuels are found close to the surface and are easy and cheap to extract. Where the coal seams or oil fields are deep underground and are fractured, the costs of exploitation are higher. If it costs more to extract the fossil fuel that it can be sold for, the resource becomes uneconomic to exploit.

Improvements in technology can open up new opportunities for energy use. Wind turbines and solar panels have become more efficient and affordable, enabling many more countries to use renewable energy.

Political factors also affect the supply of energy. Energy producing countries can group together to ensure that they receive the best price for their commodity on the world market. For example, the Organisation of the Oil Exporting Countries (OPEC), can keep oil prices high by lowering supplies when the demand for oil slumps. They sometimes also lower prices by putting more oil onto the market. Governments also sometimes subsidize energy costs to keep prices lower, in order to ensure that people can afford the energy they need in order to maintain their standard of living.

War can also impact on energy supplies. For example, during the 1990 Gulf War oil exports from the Middle East were disrupted and oil prices increased. The 2022 war in Ukraine led to a reduction in the amount of natural gas supplied to Europe by Russia, which doubled the size of energy bills in the UK. These price rises led to a change in government policy from importing less gas to generating more energy from renewable sources and nuclear power.

Local politics can also influence energy supplies. Community and environmental groups can oppose the opening of a new coal mine, the building of a wind farm or the start of fracking. Fracking, which is a technique for recovering gas and oil from shale rock, is currently banned in the UK due to opposition from these groups, who raised concerns about the large amounts of water used in the operations and the risk of triggering an earth tremor.

Government policy also influences energy supply. In order to reduce carbon emissions, the UK government has a policy of reducing the amount of fossil fuels being used. As the UK moves away from the use of oil and natural gas to renewable energy sources such as wind, there may be occasional energy shortages, leading to power cuts. Shortfalls in energy supply can occur during periods of calm weather when less electricity is produced by wind turbines. At the moment, gas fired power stations are used as a backup to ensure that the demand for electricity is met, but in the future this may not be possible. In Germany the government's decision to phase out nuclear power has also led to some energy shortages and price rises.

The impacts of energy insecurity include the increased exploration of difficult and environmentally sensitive areas, higher economic and environmental costs, lower food production, lower industrial output, and the potential for conflict where demand exceeds supply.

When energy supplies are low, the search for untapped oil or gas fields begins. Places, previously thought to be too difficult or environmentally sensitive to explore, start to be opened up. The Arctic Circle is one such area. The region already supplies the world with 10% of its oil and 25% of its natural gas. It is estimated that 22% of the Earth's undiscovered oil and natural gas reserves are located there.

Exploration of the Arctic has become easier since global warming has melted more of the ice. Also, there is no international treaty in place to protect the Arctic's fragile environment from economic development. Already several countries including Russia, Norway and the US are operating in the Arctic in spite of the higher costs involved.

Energy insecurity results in higher economic costs. Prices rise if the demand for energy increases or if there is an energy shortage. After the pandemic in 2022, gas prices rose across Europe and Asia due to the reopening of businesses. When the war in Ukraine started in 2022, supplies of natural gas from Russia were cut, further pushing up prices. To keep up supplies, Europe switched to importing liquid natural gas (LNG) from Norway, Qatar, Australia and the US. However, building LNG terminals is expensive at $10 billion each.

Energy shortages and the rise in petrol, natural gas, and electricity prices have made governments realise the importance of energy security. The transition from fossil fuels to renewables, may need to take longer to guarantee energy security and to reduce the threat of recession. During the energy transition period, more natural gas may have to be used as it is cleaner to burn than other fossil fuels like coal and oil.

Another impact of energy insecurity is increased environmental costs. When new coalmines and oil fields are opened up, the construction of new roads, pipelines and settlements for the workers can damage landscapes and habitats. When crude oil is transported by tanker there are risks of shipwrecks and oil spills. Seabirds and coasts can be polluted by the oil.

Oil can also leak from pipelines and contaminate the ground. When there was a spill from a pipeline at Huntington Beach, California in 2021, crude oil clogged beaches and wetlands and killed fish and birds. The clean-up operation took a long time.

Energy insecurity also can lead to a fall in food production. If there ar shortages of fuel and electricity, less food is produced. In LICs such a Tanzania and Mali, farmers search for firewood instead of farming leading to food shortages. If the price of biofuels rises, farmers may als switch to growing them instead of food. This has happened in Braz where sugar cane is grown to produce bioethanol, which is used instea of petrol and diesel in vehicles.

As the price of energy increases, so does the cost of manufacturin fertilizer. Food prices can rise when farmers pay more for the fertilize they use. Shortages of energy can also reduce the amount of fertilize being manufactured. This happened in Europe in 2022. Both highe prices and fertilizer shortages can result in lower crop yields and les food, particularly in low-income countries.

Another impact of energy insecurity is a fall in industrial output. I manufacturing, energy is needed as a source of power. Some industri processes simply cannot take place without electricity. Energy is also raw material. For example, oil is used in the manufacture of chemicals plastics and pharmaceuticals. Energy shortages or higher prices can bot reduce production and add to costs turning a profitable industry into loss making one. The impacts of energy insecurity are greater in som countries than in others. For example, industrial output in Pakistan i lower than it should be because of the frequent long power cuts whic occur.

There is potential for conflict in places where demand for energ exceeds supply. Energy shortages, power cuts, price rises and rationin can all cause social unrest. In many low-income countries in Africa, suc as Somalia and Sudan, many households have no access to electricity a the countries have insufficient generating capacity to meet deman Sudan shares an oil and gas field with its neighbour South Sudan, an there have been frequent disputes over its exploitation.

Different strategies can be used to increase energy supply. One strategy is to use more energy from renewable sources and another is to use non-renewable sources such as oil and natural gas.

Renewable energy sources are not diminished when they are used as they recur and cannot be exhausted. Renewable energy uses the forces of nature including moving water and air, heat from the ground and light from the Sun to make power. The types of renewable energy include biomass, wind, hydro, tidal, geothermal, wave, and solar power.

Biomass energy comes directly from organic materials and indirectly from the Sun. Plants absorb the Sun's energy through photosynthesis, and convert carbon dioxide and water into carbohydrates which provide energy for all plant growth. The energy stored in organic matter such as wood pellets, wood chips, bioenergy crops, and agricultural and domestic waste, can be burnt in power stations to heat water. The steam that is produced turns the turbines to generate electricity. Biomass can also be processed into biogas and biofuels such as ethanol and methane.

The advantage of using biomass power is that it is renewable, reliable and is not dependent on the weather. However, biomass releases carbon dioxide when burnt, but the amount is considerably less than fossil fuels.

Wind energy is electrical energy obtained from harnessing the wind with windmills or wind turbines. The wind turns the turbines to generate electricity which goes through a transformer before joining the National Grid. Wind turbines can be stand-alone or clustered together in a wind farm.

The advantages of using wind power are that it is renewable, relatively cheap and quick to install. The technology is reliable and can be used to generate large amounts of energy. There are disadvantages to using wind power – electricity is only generated if the wind is blowing at a fast enough speed. On calm days no power is produced at all. Also, wind turbines, particularly those offshore, have a life span of only 20 years even with regular maintenance, after which they have to be replaced.

The sight and sound of large onshore wind farms can have a negative impact on an area although some wind farms become tourist attractions.

Hydro-electric power is electricity generated by turbines that are driven by moving water. Hydro-electric power stations use the force of moving water to turn the turbines to generate electricity. A river is dammed and a reservoir created. Water from the reservoir flows through the turbines and into the river or a second reservoir below the dam. Water can be pumped from the lower reservoir into the upper, storing energy for use at a later time. Hydro power is used to generate 7% of the world's electricity.

The advantage of using hydro power is that it is renewable, cheap and is scaleable. Reservoirs can be used to control flooding, to provide irrigation water and for fishing and recreation. However, building dams is expensive. If a dam fails it can cause death and destruction downstream. Also, the building of large reservoirs can displace communities, destroy habitats, and disrupt the natural flow of rivers.

Tidal power stations create electricity by using the energy generated by the natural rise and fall of the tides, which are caused by the gravitational pull of the Sun and the Moon. Coastal areas experience two high and two low tides every 24 hours and 50 minutes. High tides occur 12 hours and 25 minutes apart. To generate tidal power, a barrage or dam is built across an estuary with a tidal range of at least 5 metres difference between the high and low tides.

The advantages of tidal power are that it is clean, renewable and reliable. Once a tidal-power system has been built, they are cheap to run and maintain. Tidal power does work and can supply large amounts of energy. The main disadvantages of tidal power are that tidal power stations are expensive to build, suitable locations are hard to find, and there are negative impacts on marine ecosystems. In addition, electricity can only be generated twice a day.

Geothermal power produces energy using the internal heat of the Earth's crust. Geothermal power stations are located in Iceland and other volcanically and tectonically active places. Wells are drilled into the ground to a depth of up to 10 km. The heat is drawn up to the surface in the form of water and steam. The hot water can be used to heat homes and buildings. The steam can be used to turn the turbines to generate electricity in geothermal power plants.

The advantages of geothermal power are that it is clean, renewable and reliable. The disadvantages of geothermal energy are the high construction costs, the lack of suitable locations and the possibility of contaminating groundwater.

Wave power uses the energy of waves to generate electricity. However, although there have been various trial projects, currently wave power is not being used on a wide scale. The advantages of wave energy are that it is renewable, environmentally friendly, and widely available. The disadvantages are that wave power stations are expensive to build, cause noise and visual pollution, and are unreliable.

Solar energy is the Sun's energy exploited by solar panels, collectors or cells to heat water or air or to generate electricity. Solar power is energy derived from the sun. There are two types of solar energy – solar photovoltaics and solar thermal. Solar photovoltaic technology converts sunlight into direct current electricity by using semiconductors. Solar thermal uses the heat energy from the Sun for heating or to produce electricity.

The advantages of solar power are that it is renewable, cheap to maintain and produces no pollution. The disadvantages of solar energy are that solar farms can be expensive to build. Africa has great potential for solar power but the continent currently only has 1% of global solar generation capacity because of the high cost of installing solar panels. Solar farms only operate during daylight hours and they cover a large area of ground. In addition, the life span of a solar panel is 25 years, after which they have to be replaced adding to the cost of using solar power.

Different strategies can also be used to increase energy supply from non-renewable sources of energy such as fossil fuels and nuclear power. Fossil fuels include coal, oil and natural gas. For centuries, industry and households have relied on the use of fossil fuels for heating, transport and manufacturing.

The main disadvantages of fossil fuels are that they are non-renewable, will eventually run out, and they release carbon dioxide into the atmosphere when burnt. However, until the transition to using just renewable sources of energy has been completed, consumers will continue to rely on fossil fuels in some way, shape or form.

It is estimated that there is enough coal in the world to last another 400 years. Some countries, for example the UK, have almost stopped using coal as a form of energy but other nations with abundant supplies, such as China and India, are likely to continue to use coal for many years.

The world relies on petroleum products to drive vehicles, heat buildings, and to produce electricity. In the petrochemical industry oil is used as a raw material to make products such as plastics, polyurethane, solvents, and hundreds of other intermediate and end-user goods. It is estimated that the world has about 47 years of oil reserves left if current consumption levels are maintained.

Natural gas is an important source of energy. It is thought that the world has enough gas to last over 50 years. Over 40% of global reserves are located in Iran and the rest of the Middle East.

The extraction of fossil fuels has both advantages and disadvantages. For example, the UK's oil and gas industry in the North Sea began in the 1970s and is still in operation. The Gannet Complex is a fixed drilling and production platform in the Central North Sea, 112 miles east of Aberdeen in water 95 metres deep. It is operated by Shell UK nd serves the Gannet A, B, C, D, F and G oil and gas fields. Oil from he fields is exported via the Norpipe Export System to Teesside. Gas is xported via the Fulmar gas pipeline to St Fergus.

The UK has benefitted from the extraction of North Sea oil and gas. he oil companies have paid the UK government over £334 billion in ax since the 1970s. These tax revenues have been used by the overnment to pay for public services and infrastructure.

Many highly skilled well-paid jobs in engineering, mining and eoscience have been created by the industry. In 2019, over 270,000 eople were employed directly or indirectly in the UK oil and gas ndustry. There has been a positive multiplier effect for many parts of he Eastern coast of the UK as industries develop platforms, materials nd parts for the rigs out in the sea. The economy of towns like berdeen have grown as the result of the influx of new workers and the igher incomes they earn. Even so, the North Sea oil and gas industry as a finite future and will eventually shut down as the oil and gas fields ecome exhausted and demand for fossil fuels becomes too low for the ndustry to be economic.

North Sea gas is a reliable source of energy, and supplies 60% of the JK's needs making the UK less dependent on gas imports. Not only oes North Sea gas add to the UK's energy security, it has a smaller arbon footprint than imported gas because it has travelled a shorter istance. The UK also exports gas to Europe which brings in foreign urrency.

Crude oil from the North Sea has a lower sulphur content than Middle astern oil. This makes it cheaper to process and more expensive to buy han oil from other places. It is estimated that 80% of North Sea oil is xported, earning more foreign currency for the UK.

However, there have been disadvantages involved with extracting oil nd gas from the North Sea. Large amounts of investment were needed o develop the technology required to drill through the sea bed. Over time though the cost of production has fallen from $30 a barrel to around $12, making the North Sea a more attractive place to invest.

Extracting oil in rough seas is a dangerous business. In July 1988, a series of explosions ripped through the Piper Alpha drilling platform which was located 120 miles off the north-east coast of Scotland, killing 167 people. In 2016, a helicopter taking offshore workers to an oil rig crashed off the coast of Norway with 13 fatalities.

Environmental damage has also occurred. Oil leaks have happened. In 2011 Shell UK was fined £22,500 after 200 tonnes of oil leaked from an underground pipeline at the Gannet Alpha platform. Fortunately, the oil was dispersed by the waves and did not reach shore.

Nuclear power uses the energy from the core of an atom. To release the energy, atoms are split in reactors which are powered by uranium, a finite resource. The heat produced by this nuclear reaction changes water into steam, which is used to turn the turbines and generate electricity. Nuclear power is considered to be environmentally sustainable because nuclear power stations emit low amounts of carbon.

The advantages of nuclear energy are that the reactors operate continuously producing a reliable supply of electricity. Nuclear plants can range in size from small to very large. It is hoped that small modular reactors will provide a quarter of Britain's electricity by 2050.

A disadvantage of nuclear energy is the high costs involved in building and running a nuclear power station – it is estimated that Sizewell C, a new nuclear plant in Suffolk, is likely to cost £30 billion to build. Another drawback is that at the end of their productive life, nuclear power stations have to be decommissioned. This involves the removal of nuclear material and the restoration of the site – a process that can take up to 30 years. Although the use of nuclear power has been declining in recent years, over concerns about the storage of radioactive waste and the damage caused by nuclear meltdowns, it is likely that nuclear power will continue to be an important source of electricity well into the future.

All around the world, countries are moving towards a future when the use of energy resources is sustainable. Sustainable energy is when energy supply matches demand, when the production and consumption of energy does not contribute to climate change, and when the current need for energy does not limit the ability of future generations to meet their own needs for power. Sustainable energy can be achieved through a combination of energy savings, energy efficiency measures and technologies, as well as the use of renewable energy sources.

Energy conservation is an important part of achieving a sustainable energy future. Energy conservation involves using and wasting less energy. Individuals can reduce the amount of energy they use and decrease the size of their carbon footprint by making small changes to their lifestyles. For example, people can reduce their use of energy by walking, cycling, using public transport, or by driving an electric vehicle. In their homes, people can use a renewable source of energy such as solar panels, insulate their lofts and recycle their waste. If every UK household lowered the amount of energy they use by 4 tonnes, this could reduce the UK's carbon emissions by 80% by 2050.

Homes can be designed so that they use less energy for heating, cooling and lighting while still staying warm in winter and cool in summer. One way to do this is to improve home insulation. Better insulation can be achieved by lagging hot water tanks, and draught proofing windows, doors and letterboxes. Fitting cavity wall and loft insulation also helps prevent heat loss.

Another way to conserve energy is to use domestic appliances which use less electricity and are more energy-efficient. Gas boilers which use less gas can be installed or replaced with heat source pumps. By adopting these measures, the average UK household could reduce its carbon emissions and save money.

Workplaces can be designed with energy conservation in mind. For example, solar panels can be installed on roofs so that more renewable energy is used, and offices and factories can be better insulated to prevent energy loss. Electric vehicle charging points and cycle racks can be installed so that workers choose to travel to work by bike or electric car.

Transport can become more sustainable with lower carbon emissions if more energy efficient vehicles are used. For example, older cars can be replaced by newer models which use less petrol and diesel, or with electric vehicles. Electric trains can be used instead of those that run on diesel, and buses can switch to using biofuels. Modern aircraft are already fuel efficient.

Demand reduction, which means encouraging people to use less energy, can also play its part. Governments can change building regulations so that homes become more energy efficient. Information campaigns can be run to raise awareness of energy conservation.

The use of technology to increase the efficiency of burning fossil fuels is becoming a reality. Older types of light bulb, which use a lot of electricity, are being replaced by more efficient LED bulbs. Smart devices such as thermostats, radiator controls and meters are being installed in homes to reduce energy wastage by automatically adjusting to the changing energy needs of the household.

An example of a local renewable energy scheme which provides sustainable supplies of energy can be found in Zimbabwe, a lower-middle income country. There is a renewable energy scheme run by a technology and utility startup business called NeedEnergy. The company generates a clean renewable supply of solar energy which it sells to local businesses, some of which are located in Ridgeway North, a boutique shopping mall in the affluent suburb of Harare. The mall is not connected to the national grid; instead, it relies on the guaranteed supply of electricity provided by NeedEnergy generated by their solar panels.

As solar power is only generated when the Sun is shining, the company uses artificial intelligence (AI) to analyse weather data and forecasts in order to estimate how much electricity will be produced each day. They also collect data from their customers' smart meters and use AI to predict how much electricity their customers will need.

This local renewable energy scheme not only has a lower carbon footprint but also lower maintenance and running costs. The electricity provided by NeedEnergy is more reliable and cheaper than using the national grid or back-up diesel generators. The small businesses in the mall now have energy security and there is less air pollution and lower carbon emissions. This enterprise is sustainable because it meets the needs of the present generation without compromising the ability of future generations to meet their own.

Bibliography

General References

Boulter, Maliphant, Plowman, Ryan *GCSE AQA Geography For the Grade 9 -1 Course The Revision Course*
(https://www.cgpbooks.co.uk/secondary-books/gcse/humanities/geography)

Ross, Rowles, Holmes, Digby, *GCSE Geography AQA* (Oxford University Press 2016)

Useful Websites

AQA
https://filestore.aqa.org.uk/resources/geography/specifications/AQA-8035-SP-2016.PDF
Arctic National Wildlife Refuge
https://www.fws.gov/refuge/arctic

BBC https://www.bbc.co.uk/
Birmingham https://www.birmingham.gov.uk/
Bristol https://www.bristol.gov.uk/
British Geological Survey https://www.bgs.ac.uk/discovering-geology/
British Red Cross https://www.redcross.org.uk/
Cadair Idris https://www.visitwales.com/things-do/adventure-and-activities/walking/guide-to-walking-up-cader-idris
City of London https://www.cityoflondon.gov.uk/things-to-do/green-spaces/epping-forest
Cockermouth https://www.gov.uk/government/news/environment-agency-completes-emergency-flood-works-in-cockermouth
Cost estimation for coastal protection – summary of evidence https://assets.publishing.service.gov.uk/media/6034ee168fa8f5432c277c23/Cost_estimation_for_coastal_protection.pdf
Department for Transport https://www.gov.uk/government/organisations/department-for-transport
Fairtrade https://www.fairtrade.org.uk/
Forest Stewardship Council International https://fsc.org/en
International Monetary Fund https://www.imf.org/en/Home
London https://www.london.gov.uk/
Met Office https://www.metoffice.gov.uk/
Minehead Coastal Defence Scheme https://www.bbc.co.uk/bitesize/guides/z38dv4j/revision/6
National Oceanographic and Atmospheric Administration https://www.noaa.gov/
Newcastle https://www.newcastle.gov.uk/
Office for National Statistics https://www.ons.gov.uk/

River Rheidol https://www.discoverceredigion.wales/areas-of-ceredigion/ceredigions-river-valley-routes/rheidol-valley/
Secretariat of the Antarctic Treaty https://www.ats.aq/index_e.htm
Shell Petroleum Development Company of Nigeria (SPDC) https://www.shell.com.ng/about-us/what-we-do/spdc.html
Snowdonia National Park https://snowdonia.gov.wales/
Somerset Levels and Moors: reducing the risk of flooding https://www.gov.uk/government/publications/somerset-levels-and-moors-reducing-the-risk-of-flooding/somerset-levels-and-moors-reducing-the-risk-of-flooding
Somerset Rivers Authority https://www.somersetriversauthority.org.uk/
Teachit Geography https://www.teachit.co.uk/geography
The World Factbook https://www.cia.gov/the-world-factbook/
UK Aid direct https://www.ukaiddirect.org/
UNESCO World Heritage Sites https://whc.unesco.org/en/list/
UNICEF https://www.unicef.org.uk/
United Nations Convention to Combat Desertification – The Great Green Wall Initiative https://www.unccd.int/our-work/ggwi
United Nations Framework Convention on Climate Change (COP27) https://www.un.org/en/climatechange/cop27
United Nations https://www.un.org/en/
West Kirby Flood Alleviation Scheme https://www.wirral.gov.uk/environmental-problems/coastal-protection/west-kirby-flooding
World Bank https://data.worldbank.org/indicator/NY.GNP.PCAP.CD
World Food Programme https://www.wfp.org/
World Wildlife Fund https://www.wwf.org.uk/
Zoological Society of London https://www.zsl.org/

Index

About the Author

Elizabeth Paice is an AQA GCSE Geography Examiner and Geography Teacher with considerable teaching and examining experience.

She studied Geography at University College London, gaining a BA Hons in Geography, followed by PGCE at Jesus College Oxford. She has an M.A. Education from the Open University and an NPQH from the Institute of Education, London.

She is a Governor of a secondary school in west London and a contributor to Teachit Geography.

To contact Elizabeth, visit her website:

www.lizpaice.com

Printed in Great Britain
by Amazon